Thomas Kirkland

Elementary Statics

Principally designed for the use of candidates for the first and second class

certificates, and for the intermediate examinations

Thomas Kirkland

Elementary Statics
Principally designed for the use of candidates for the first and second class certificates, and for the intermediate examinations

ISBN/EAN: 9783337179304

Printed in Europe, USA, Canada, Australia, Japan

Cover: Foto ©Andreas Hilbeck / pixelio.de

More available books at **www.hansebooks.com**

Miller & Co's Educational Series.

ELEMENTARY STATICS,

Principally designed for the use of Candidates for First and Second Class Certificates, and for the Intermediate Examination.

WITH NUMEROUS EXAMPLES AND EXERCISES

BY

THOMAS KIRKLAND, M.A.,

Science Master, Normal School, Toronto.

FOURTH EDITION.

TORONTO:
ADAM MILLER & CO.
1878.

Entered according to Act of the Parliament of Canada, in the year one thousand eight hundred and seventy-seven, by ADAM MILLER & Co., in the office of the Minister of Agriculture.

PREFACE.

THE following pages contain the substance of the lectures on Statics, delivered to the students of the Second Division in the Normal School, Toronto, during the past five years. They are now published in compliance with resolutions passed at several Teachers' Associations, and at the request of many excellent teachers in High and Public Schools.

Throughout the book, I have constantly kept in view the requirements of Candidates for Second Class Certificates and for the Intermediate Examination, who have not, in general, obtained the knowledge of mathematics, which most elementary books on Statics presuppose. In the perusal of the following pages, which will be found to contain all that is usually comprised in Elementary Treatises on Statics, the only thing required of the student is a competent knowledge of the First Book of Euclid and simple equations in Algebra.

Each chapter is divided into sections with the same fundamental idea running through the section. The student will not, therefore, be perplexed by trying to learn too much at once, but will be obliged to fix his attention on one subject at a time.

PREFACE.

He will thus the more easily master the difficulties it presents. Each section contains one or more examples fully worked out; and amongst the answers, at the end of the book, very full hints are given for the solution of all the more difficult problems. These features, it is hoped, may be of service to those who study the book without the aid of a teacher. To each section is appended a collection of questions and problems designed to test the student's knowledge of the subject-matter of the section, to awaken activity of thought, and to exercise his invention in the solution of problems by the application of the principles contained in the section. Great pains have been taken in constructing and selecting the questions for the different exercises. The object has been not to select intricate and puzzling questions, but such as will serve to determine from the form of solution, whether the student has grasped the fundamental principles of the particular subject, and is capable of applying them. As a guide to beginners the more important propositions have been printed in black-letter type, the less important in italics.

Throughout the work I have endeavored to explain as clearly as I could the leading ideas of Elementary Statics and to get rid of all difficulties that are not inherent in the subject itself. But it may be well to remind the student that after all that is possible has been done in the way of exposition and illustration, the subject will still present difficulties to beginners—difficulties which can only be overcome by the labor of close thinking.

PREFACE.

Students unfamiliar with geometrical deductions are recommended, at the first reading, to omit the following:

Exercise III., Chapter II., and Sections II. and III., Chapter III.; Sections I. (except the Introduction), III., and Exercise IV., Chapter VI.; Section III., Chapter VII.

After reading the chapter on the Mechanical Powers, the student will be better able to master the omitted sections. By this division into sections, it is hoped that the book has been adapted to the requirements of beginners as well as to the wants of advanced students.

I have to tender my thanks to several friends for suggestions and assistance which have been of the greatest service to me, and particularly to Professor Young for suggesting several important improvements in the work, and for the excellent collection of Examination Papers in Chapter XII., which add much to the value of the book.

Any remarks on the work, and especially indications of errors, omissions, or difficulties will be thankfully received.

It will give me much pleasure if these pages shall in any way contribute to advance the taste for a science that is at once useful and interesting.

THOMAS KIRKLAND.

Normal School, April 1877.

CONTENTS.

CHAPTER I.

	PAGE.
Definitions and Preliminary Notions.	1

CHAPTER II.

Parallelogram of Forces.

SECT. I.—Experimental Proof and Examples of Forces acting at right angles to each other 7

SECT. II.—Forces acting at angles of 60°, 30°, 45°, and their supplements. 13

SECT. III.—Questions requiring for their solution a knowledge of easy deductions from the First Book of Euclid 17

CHAPTER III.

The Triangle and Polygon of Forces.

SECT. I.—Proof of Triangle of Forces and Easy Exercises 19

SECT. II.—Extension of the Principle of the Triangle of Forces, and its application to find the tension of strings 22

SECT. III.—The reaction of hinges and smooth surfaces, and three forces keeping a body at rest 25

CHAPTER IV.

The Resolution of Forces.

SECT. I.—Method of resolving a given force into two others which act in given directions 30

SECT. II.—Conditions of Equilibrium when any number of forces act at a point 36

CHAPTER V.

Parallel Forces.

CONTENTS.

CHAPTER VI.
Moments of Forces.

PAGE.

SECT. I—Principle of Moments. Equilibrium of a body capable of turning round a fixed point. Equilibrium of a body acted on by any number of forces in one plane. Hints for the solution of problems.................... 48

SECT. II.—The application of the Principle of Moments to the Lever.................................... 57

SECT. III.—Equilibrium of a body acted on by any number of Forces...................................... 63

CHAPTER VII.
The Centre of Gravity.

SECT. I.—Centre of Gravity of a uniform straight rod, and of bodies lying in the same straight line............... 69
SECT. II.—Properties of the Centre of Gravity 74
SECT. III.—Centre of Gravity of plane areas., &c......... 78

CHAPTER VIII.
Mechanical Powers.

SECT. I.—The Lever, Balances, &c..................... 89
SECT. II.—The Wheel and Axle....................... 98
SECT. III.—The Pulley............................. 100
SECT. IV.—The Inclined Plane....................... 112
SECT. V.—The Wedge.............................. 119
SECT. VI.—The Screw.............................. 121

CHAPTER IX
Virtual Velocities as Applied to Machines.

CHAPTER X.
Friction.

CHAPTER XI.
The Application of Similar Triangles to the Solution of Statical Problems...................... 140

CHAPTER XII.
Examination Papers set to Candidates for First and Second Class Certificates, from 1871 to 1876 inclusive........ 146

APPENDIX.
Proof of the Parallelogram of Forces.
Results, Hints, &c., for the Exercises.

CHAPTER I.

DEFINITIONS AND PRELIMINARY NOTIONS.

1. Mechanics. The science of *mechanics* treats of the effects of *force;* we must then begin by explaining what we mean by the term *force*.

2. *Def. of Force*. Any cause which moves or tends to move, a body or which changes or tends to change its motion, is called force.

3. In order to conceive the existence of a force, we must conceive that there is something upon which it can act and which may be called matter. A limited portion of matter is called a *body*. When the body is so small that for the purpose of any discussion the relative position of its parts need not be considered, it is called a *material particle*.

4. **Def.** A *material particle* may, therefore, be defined as a portion of matter occupying an indefinitely small space. It is spoken of for shortness as a *particle*.

5. Rigid Body. *A rigid body is one in which the relative position of the particles cannot be altered by the action of any finite force.*

6. **Equilibrium.** When several forces act simultaneously upon a body, their individual tendencies to produce motion may so counteract each other that no motion will ensue; these forces are then said to be in *equilibrium*.

7. **Statics.** Statics is the science that investigates the relations existing among forces in maintaining rest or preventing change of motion; it is the Science of Equilibrium.

8. **Magnitude of a Force.** The first thing to be done in order to bring force under mathematical treatment, is to find some means by which the magnitudes of

different forces may be numerically represented. In order to do this we must fix, by common consent, upon some standard force which, producing a known effect, may be taken as a unit of force. Nothing can be more convenient for this purpose than a given *weight*. Take, for instance, a weight of 1lb. If a force, when acting vertically upwards, will just sustain this weight, the force may properly be described as a force of 1lb. If it will just sustain two such weights fastened together, it may be described as a force of 2lbs.; and so on. And generally, if a force is just capable of sustaining a weight of P lbs., it is called a force of P lbs., or more briefly a force P; and a force just capable of sustaining a weight of Q lbs. is called a force Q. Whenever, therefore, a force is denoted by a letter of the alphabet, as P,Q,R, &c., it will be understood that these letters represent numbers, and that these numbers are the number of pounds which the forces are just capable of sustaining.

9. **Force measured in terms of weight for convenience.** The student will understand that it is simply for convenience, and not of necessity, that reference is made to *weight* as a measure of force. Instead of taking as the unit of force that force which will just sustain a weight of 1lb., a force might be taken that would just bend a given spring, but this would not be so convenient a method.

10. **Representation of Forces by Lines.** The three elements specifying a force, or the three elements which must be known, before a clear notion of the force under consideration can be formed, are:

    1. Its point of application.
    2. Its direction.
    3. Its magnitude.

Hence we may conveniently represent forces by straight lines. For we may draw a straight line from the point of application of the force in the direction of the force, and containing as many units of length as the given force contains units of weight. An example will make this plainer.

Ex. Suppose a force of 5lbs., inclined at an angle of 30° to the horizon, to be acting at the centre of a horizontal rod.

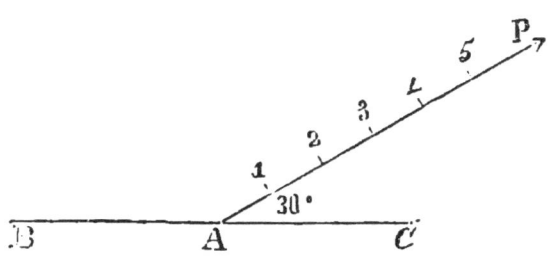

Let BAC be the rod, A the centre; draw at A a straight line making an angle of 30° with the horizon. Take a portion of this line AP containing five units of length, that is, as many units of length as there are units of weight in the given force. Then we say that AP represents the given force in every necessary particular.

In magnitude, by the number of units in its length.

In direction, by its inclination to the horizon viz., 30°.

In point of application, viz. at A the centre of the rod.

It is evident that *any* force may in this manner be fully represented by a straight line; and this mode of representing forces by lines has the following advantages :—

1. The three elements specifying a force are all presented to the eye at once.

2. It simplifies the enunciation of many theorems in mechanics.

3. It enables us to infer mechanical propositions from corresponding propositions in geometry.

11. **Distinction between the force AP and the force PA.** It may be inferred from the preceding article that a force AP applied at A has not the same effect as a force PA applied at the same point; for a

force AP would tend to move the body at A in the line AP towards P, but a force PA would tend to move the body at A in an opposite direction. The two forces are equal in magnitude and act along the same straight line, but tend to produce motions in opposite directions. The "force AP" is generally used to indicate the force represented in magnitude and direction by the line AP, acting in a direction from A towards P.

12. **Ways in which force may be exerted.** A force may be exerted in various ways, but only the following three will come under our notice.

1. **Pressures.** We may push a body by the hand or by a rod; a force so exerted is called a *pressure*.

2. **Tensions.** We may pull a body by means of a string or by means of a rod; the force exerted by means of the string, or by means of the rod used like a string, is called a *tension*.

3. **Attractions.** The earth exerts a force tending to draw bodies towards its centre; a magnet exerts a force on certain bodies brought under its influence; such forces are called *attractions*.

13. **Gravity.** All bodies if left to themselves would fall towards the centre of the earth, and the force which produces this tendency is called the *force of gravity*. This force is generally of different magnitudes for different bodies, varying in some degree with their size and substance, but it is found to be constantly of the same magnitude for the same body at the same place on the earth's surface; but acts with slightly different force at different parts of the earth's surface. Its direction is always perpendicular to the surface of still water, and is usually designated as the *vertical line*. A plane perpendicular to this line is said to be *horizontal*.

If a body be prevented from falling by the interposition of some object such as a hand or a table, the body exerts a pressure on the hand or table. Hence we obtain the following definition.

14. **Weight.** *The precise amount of pressure which any body at rest exerts in the direction of the earth is called the weight of that body.*

15. **The Transmissibility of Force.** If a weight be attached by a cord to a spring balance, and the weight of the cord be neglected, the effect will be the same at whatever point in the cord the weight is attached. The same would be the case if the cord were passed over a perfectly smooth peg before being attached to the balance. Similarly, a force may be applied to a body by means of a rigid rod, and if the rod be supported independently, the result will be the same. The general principle here illustrated, and which we derive from experiment and observation, may be stated as follows :—

Principle of the Transmissibility of Force. **When a force acts on a rigid body the effect of the force will be unchanged at whatever point of its direction it be applied, provided, this point be a point of the body, or be rigidly connected with the body.**

16. **Object of Statics.** It is the principal object of Statics to investigate what must be the relations between a given set of forces, as regards their magnitude and direction, in order that they may be in equilibrium. To this end it is necessary to investigate under what circumstances and in what manner it is possible to replace a given set of forces by another of a simpler or more convenient kind without affecting the state of equilibrium.

17. **Bodies in nature not perfectly rigid.** In the following investigations we will suppose strings perfectly flexible, surfaces perfectly smooth, and bodies perfectly rigid. Now, there is no such thing in nature as a rigid body. Some are more nearly rigid than others. Iron is more rigid than wood, and wood than water. It may at first sight appear useless to make investigations upon an hypothesis which never corresponds to reality. The method of proceeding is, however,

to base the calculations upon the hypothesis of perfect rigidity, &c., and then to make such an allowance in the result, for want of this flexibility, smoothness, and rigidity, as experiment shows to be necessary for the particular substance with which we happen to be dealing. In this way our final results, though not mathematically exact, are yet in most cases sufficiently near the truth for all practical purposes.

QUESTIONS ON CHAPTER I.

1. Define Statics.
2. Define force. Mention any forces you know of, and show how your definition applies to them. Is a book lying on a table a force?
3. How is force measured in statics? Show how your measure of force applies to a force acting in any direction whatever.
4. What would you understand by a force of 10, or a force of P in the absence of other information respecting such a force?
5. Define the direction of a force.
6. Show clearly how a straight line may be made to represent a force in magnitude and direction.
7. Show the propriety of representing statical forces by straight lines, and mention any advantages which belong to this mode of representation.
8. If a force which can just sustain a weight of 5lbs. be represented by a straight line whose length is 1 ft. 3 in., what force will be represented by a straight line 2 ft. long?
9. When a weight hangs by a string, what is meant by the statement that the tension of the string is the same throughout? Is the tension the same if the string has weight?
10. What is meant by weight? In what direction does it act? In how far is the weight of a given substance invariable? Does 1lb. at the equator weigh the same as 1 lb. at the pole?
11. State the "Principle of Transmissibility of Force," and give illustrations.

CHAPTER II.

THE PARALLELOGRAM OF FORCES.

SECTION I.

Experimental Proof; and Examples of Forces acting at right angles to each other.

18. **Resultant and Components.** Two or more forces acting on a particle in any direction may be represented by a single force which will be equivalent to the whole system. For suppose such a system did not keep the particle at rest, it is evident that the particle would begin to move in a certain direction. A single force of the necessary magnitude and acting in a direction exactly opposite to that in which the particle would begin to move, would keep it at rest. This force would counterbalance the original set of forces, and a force equal and opposite to it would produce the same effect as the first set of forces, and is therefore called their resultant.

19. **Def. of Resultant.**—Hence the single force which represents the combined effects of two or more forces, is termed their *resultant*.

20. **Def. of Components.** Those forces which form a system equivalent to a single force are called its components.

21. **Composition of Forces.** The method of finding the resultant of two or more forces is called the *Composition of Forces*.

22. *To find the resultant of two or more forces acting on a particle along the same straight line.*

If a force of P lbs. acts upon a particle in a certain direction, and another force of Q lbs. acts upon the particle in the same direction, it is evident that the two forces together will produce the same effect as a single force of $(P + Q)$lbs. acting in the direction of the two forces. Or, if R be the numerical value of the equivalent force in lbs.,

$$P + Q = R.$$

If P act in one direction and Q in the opposite direction, and P be the greater force, we shall have

$$P - Q = R.$$

23. The Parallelogram of Forces. If, however, the forces P and Q, instead of acting in the same line, act upon the particle in two different directions, then their combined effect is the same as that of a single force, whose magnitude and direction are found by the following proposition which is called the *Parallelogram of Forces* :—

If two forces acting on a particle be represented in magnitude and direction by straight lines drawn from the particle, and a parallelogram be constructed having these straight lines as adjacent sides, then the resultant of the two forces is represented in magnitude and direction by that diagonal of the parallelogram which passes through the particle.

24. Illustration of the Parallelogram of Forces. The following illustration will help the student to understand this enunciation :—

Let P and Q be two forces acting on the particle A in the directions AP, AQ, respectively.

Take AP, AQ, each as many units of length as there are units in the forces P and Q respectively.

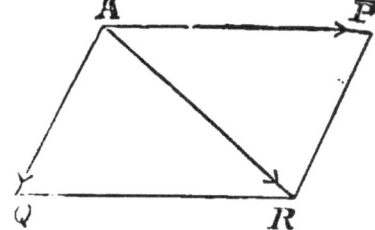

For example, if P = 5 lbs.,
and Q = 3 lbs.,
and one inch be chosen as the unit of length,
take AP = 5 inches,
AQ = 3 inches,
then the lines AP, AQ represent the forces P and Q in magnitude, direction, and point of application.

Complete the parallelogram APRQ, and draw the diagonal AR. Then the proposition states that the diagonal AR will represent the resultant of P and Q in magnitude and direction, *i.e.*, the resultant of P and Q contains as many pounds as AR contains inches. P and Q might, therefore, be removed, and the single force, represented by the line AR substituted in their stead.

25. **Experimental proof of the Parallelogram of Forces.** Let A and B be two pulleys fixed in a vertical board.

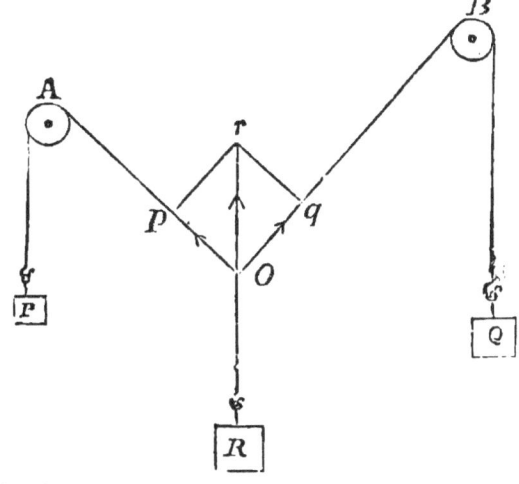

Let P and Q be two weights attached to fine cords passing over the pulleys and united at the point O. At this point let another weight be attached by a fine cord. Let the system be now abandoned to its own action and it will settle into a fixed position represented in the figure.

The proof of the Parallelogram of Forces usually given in works on Statics is unsuitable for those who are not familiar with mathematical reasoning. It is deduced from the particular case of two equal forces acting upon a particle; but as the case of two equal forces must itself be deduced from experience, there appears to be little gained by this method of proof either in point of clearness or of certainty. We have therefore inserted an experimental proof of the proposition itself, giving the usual proof in the appendix.

Forces equal to the weights P and Q now act at the point O, in the directions OA, OB, and the effect of these is counteracted by the weight R, acting vertically downwards. Therefore, the weight R is equal and opposite to the resultant of P and Q. If, now, the line O*p* be measured off in the direction OA containing as many inches as there are ounces in the weight P, and the line O*q* be measured off in the direction OB containing as many inches as there are ounces in the weight Q, and the parallelogram O*prq* be completed; then it will be found that the vertical line OR produced backwards passes through the point *r*, and that O*r* contains as many inches as there are ounces in the weight R. Hence, if two forces O*p*, O*q*, act upon a particle at O, their combined effect is equal to the effect of a single force represented in magnitude and direction by the diagonal of the parallelogram O*prq*.

Care must be taken in constructing the parallelogram of forces, that the components of the forces act *from* the angle of the parallelogram from which the diagonal is drawn.

26. **Remarks on the Experimental proof of the Parallelogram of Forces.** The principle of the parallelogram of forces may be regarded as the foundation of the science of Statics. By the means indicated in the last article it can be proved, so far as such a proposition can be proved experimentally. The more accurately the experiments are made, and the more they are varied in their circumstances, so much the more certain will they render the truth of the proposition. But the best evidence we have of its truth is of an indirect character. Direct experiment, as we have seen, affords a very strong presumption in favor of the truth of the proposition. We then assume it to be true, and construct on this basis the whole science of statics. Then we compare many of the results obtained with observation and experiment, and we find the agreement so close in every case that we may fairly infer that the principle on which the theory rests is true.

PARALLELOGRAM OF FORCES.

27. All the Forces act in one plane. All the forces with which we are concerned are supposed throughout this treatise to act in one plane, and that plane is coincident with the plane of the paper. For brevity we shall generally leave this to be understood in the investigations. Of course all the operations are supposed to be carried on in the same plane.

EXAMPLES.

1. Two forces of 4lbs. and 3lbs. respectively act on a particle at right angles to each other; find the magnitude and direction of the resultant.

Let AB, AC, represent the forces acting on the particle at the point A, $AB = 3$, and $AC = 4$.

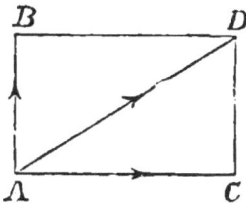

Complete the rectangle ABDC, and join AD. Then AD represents the resultant in magnitude and direction.

$$AD^2 = AC^2 + CD^2 \text{ (Euc. I. 47)}$$
$$= AC^2 + AB^2 \text{ (Euc. I. 34)}$$
$$= 4^2 + 3^2$$
$$= 25,$$
$$\therefore AD = 5.$$

2. Two forces which are to each other as 3 to 4, act on a particle at right angles to each other and produce a resultant of 15lbs.; find the forces.

Let AB, AC (last fig.) represent the forces. Then since the forces are as 3 to 4, one is three times some unit of which the other is four times.

Let x be this unit, the forces will then be $3x$ and $4x$ respectively.

$$AD^2 = AC^2 + CD^2$$
$$= AC^2 + AB^2$$
$$= (4x)^2 + (3x)^2;$$
$$\text{or } 15 = 25x^2,$$
$$\therefore x = 3.$$

And therefore the forces are 9lbs. and 12lbs.

Or thus: When $AB = 3$, and $AC = 4$, then $AD = 5$.
But in this case $AD = 15$ or 5×3.
Hence $AB = 3 \times 3$ or 9.
And $AC = 4 \times 3$ or 12 as before.

EXERCISE I.

1. Define the resultant of any number of forces. How would you find the resultant of two or more forces which act on a particle and in the same straight line?

2. Two forces of 7lbs. and 8lbs. act upon a particle in the same direction; what will be the magnitude and direction of their resultant?

3. What is the resultant of two forces measured by 3lbs. and 4lbs., respectively, acting in opposite directions?

4. Three forces of 4lbs., 2lbs., and 3oz., respectively, act upon a particle in the same direction; and in the opposite direction, forces of 8oz., and 5lbs., act. What other single force will keep the particle at rest?

5. Can the resultant of two forces in any case exceed the sum of the forces? Under what circumstances is it least? Can it be zero?

6. Can the resultant in any case be equal to one of the components? If so, what are the conditions?

7. State the Parallelogram of Forces. Explain the meaning of the terms employed in your statement; apply it to show that if four forces acting on a particle be represented by the sides of a rectangle taken in order, they will be in equilibrium.

8. Two forces of 9lbs. and 12lbs. act upon a particle in directions at right angles to each other; find the magnitude of their resultant.

9. If two forces of 5lbs. and 12lbs. act at the same point at right angles to each other, what single force will produce the same effect?

10. Two forces, one of which is three times the other, act along the adjacent sides of a square; find the resultant.

11. Two forces at right angles to each other, one of them being 8lbs.; find the other when the resultant is 10lbs.

12. If a weight be supported by two strings tied to it which are pulled in directions at right angles to each other, by forces of 18lbs. and 24lbs. respectively; find the weight.

13. Three cords are tied together at a point; one of these is pulled in a northerly direction with a force of 6lbs., and another in an easterly direction with a force of 8lbs. With what force must the third force be pulled in order to keep the whole at rest?

14. Which will be the more effective, two men pulling with a single rope, the strength of each being 2P, or two men pulling with two ropes, at an angle of 90°, the strength of each being 3P?

15. Two forces which act at right angles on a particle have the ratio of 9 to 40, and their resultant is 123lbs.; find the magnitude of the forces.

16. Two forces whose magnitudes are as 3 to 4, act on a particle in directions at right angles to each other, and produce a resultant of 2lbs.; find the forces.

17. The smaller of two forces which act at right angles is 7.2lbs., and the sum of the resultant and the larger force is 259.2lbs.; find the resultant and the larger force.

18. What are the conditions of equilibrium when two forces act on a particle?

SECTION II.

Forces acting at angles of 60°, 30°, 45°, and their supplements.

Given certain angles and one side of a right-angled triangle, to find the other two sides.

28. By means of the parallelogram of forces we may always find the resultant of two forces acting on a particle when we know the magnitude of the forces and the angle between them, but the magnitude of the resultant

can be calculated numerically without the aid of trigonometry, only in the case of a few angles, such as 30°, 60°, 45°, and their supplements.*

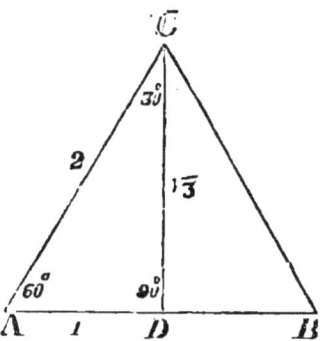

1. Let ABC be an equilateral triangle, and therefore each of its equal angles 60°. From C draw CD perpendicular to AB. AB is bisected in the point D.

Since CAD = 60°, and ADC = 90°, ACD is therefore = 30°.

Let AC = 2, then AD = 1.
And $AC^2 = AD^2 + CD^2$;
or $2^2 = 1^2 + CD^2$;
therefore $CD^2 = 3$,
or $CD = \sqrt{3}$.

Hence in any right-angled triangle whose acute angles are 30° and 60° respectively, the side opposite to the 30° is half the hypothenuse, and the side opposite the 60° is half the hypothenuse, multiplied by $\sqrt{3}$.

Since the sides of the above triangle are in the ratios 2, 1, $\sqrt{3}$ if one side be given the other two may easily be found. Thus

If AC = 5, then AD = $\frac{5}{2}$, and CD = $\frac{5}{2}\sqrt{3}$.
" AD = 5, " AC = 10, " CD = $5\sqrt{3}$.

If CD be the given side, divide each of the ratios by $\sqrt{3}$ then,

---

* The supplement of an angle is the quantity by which the angle falls short of 180°. The supplement of 60° is 120°.

if $CD = 1$, $AC = \dfrac{2}{\sqrt{3}}$ and $AD = \dfrac{1}{\sqrt{3}}$.

„ $CD = 5$, $AC = \dfrac{10}{\sqrt{3}}$ and $AD = \dfrac{5}{\sqrt{3}}$.

The student is recommended always to make a triangle having its sides in the above ratios; then having given any one of the sides, he can readily find the other two. A similar remark applies to the triangle ABC below.

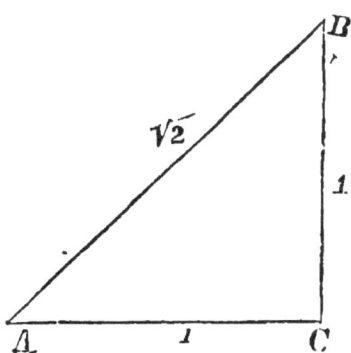

Let ACB be a right-angled isosceles triangle having $AC = CB = 1.$; then $AB = \sqrt{2}$.

If AC or $CB = 5$, then $AB = 5\sqrt{2}$.

If $AB = 5$, to find AC or CB, divide the ratios of the sides by $\sqrt{2}$, then,

If $AB = 1$, AC or $CB = \dfrac{1}{\sqrt{2}}$;

and if $AB = 5$, „ $\dfrac{5}{\sqrt{2}}$.

EXAMPLES.

1. To find the resultant of two forces of 12lbs. and 8lbs. respectively, acting on a particle at an angle of 60°.

Let $AC = 8$, $AB = 12$, and the angle $BAC = 60°$; it is required to find the resultant AD.

Produce AB, and from D draw DE perpendicular to AB produced.

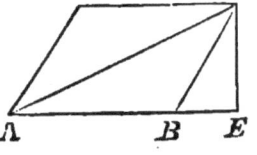

Then $BD = AC = 8$, and $DBE = CAB = 60°$ (Euc. 1.29), and therefore $BDE = 30°$.

Since $BD = 8$, then $BE = 4$, and $DE = 4\sqrt{3}$ (Art. 28);

and $AD^2 = AE^2 + DE^2$
$= (12 + 4)^2 + (4\sqrt{3})^2$
$= 304 = 16 \times 19$;
$\therefore AD = 4\sqrt{19}$.

2. Two forces of 10 lbs. and 42 lbs., act on a particle at an angle of 120°; find their resultant.

Let $AB = 10$, $AC = 42$, and the angle $BAC = 120°$.

Since the angles BAC, ACD are together equal to 180°, and $BAC = 120°$, then $ACD = 60°$, (Euc. 1.29), and therefore $CDE = 30°$.

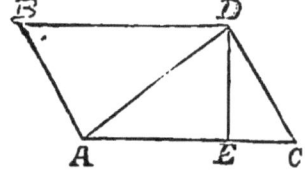

Also, since $CD = AB = 10$, then $CE = 5$, and $DE = 5\sqrt{3}$ (Art. 28.);

and $AD^2 = AE^2 + ED^2$
$= (42 - 5)^2 + (5\sqrt{3})^2$
$= 1444$;
$\therefore AD = 38$.

EXERCISE II.

1. Two forces of 9lbs. and 56lbs. act upon a particle at an angle of 60°: find their resultant.
2. The directions of two forces represented by 3lbs. and 5lbs. respectively, include an angle of 60°: find the magnitude of their resultant.
3. Two forces, whose magnitudes are represented by the numbers 2 and $3\sqrt{3}$, act upon a particle at an angle of 30°: find their resultant.

4. Two forces of 45lbs. and 325lbs. act upon a particle at an angle of 120°: find their resultant.

5. Two rafters making an angle of 60°, support a chandelier weighing 90lbs.; what will be the pressure along each rafter?

6. Two forces of 2lbs. and 3lbs. respectively, act at an angle of 45°: find their resultant.

7. Forces of 17lbs. and $24\sqrt{3}$lbs. act on a particle at an angle of 135°: find the magnitude of their resultant.

8. Show that the resultant of the forces 7 and 14 acting at an angle of 120°, is the same as the resultant of the forces 7 and 7 acting at an angle of 60°.

9. Three posts are placed in the ground so as to form an equilateral triangle, and an elastic string is stretched round them, the tension of which is 6lbs.: find the pressure on each post.

10. Three pegs A, B, C, are stuck in a wall in the angles of an equilateral triangle, A being the highest and BC being horizontal; a string, the length of which is equal to four times a side of the triangle, is hung over them, and its two ends attached to a weight W: find the pressure on each peg.

SECTION III.

Questions requiring for their solution a knowledge of easy deductions from the propositions of Euclid, Book I.

EXERCISE III.

1. The resultant is always nearer to the greater force.

2. The greater the angle between two forces the less is their resultant.

3. If the angle between two equal forces acting on a particle be 120°, what is their resultant?

4. A string passing round a smooth peg is pulled at each end by a force equal to the strain upon the peg; find the angle between the two parts of the string.

5. Show that the resultant of two forces, which act on a

particle in directions not in the same straight line, must be less than the sum of the two forces, and greater than their difference.

6. Two forces act on a particle at right angles to each other, and the resultant is double of the less; show that the angle which the resultant makes with the less is double the angle which it makes with the greater.

7. Let ABC be a triangle, and D the middle point of BC; if the three forces represented in magnitude and direction by AB, AC, DA, act upon the point A, find the magnitude and direction of the resultant.

8. ABCD is a square, a force of 1lb. acts along the side AB from A to B: a force of 1lb acts along the side AD from A to D; and a force of 2lbs. acts along the side CB from C to B; determine the magnitude and position of the resultant of the three forces.

9. AB and AC are adjacent sides of a parallelogram, and AD a diagonal; AB is bisected in E: show that the resultant of the forces represented by AD and AC is double the resultant of the forces represented by AE and AC.

10. Two strings, each bearing a weight of 6lbs. pass over two pulleys A and B, lying in the same horizontal line, and 20 inches apart; the ends of the cords are joined at C to a third weight: what must that weight be that C may be exactly 10 inches below the line AB?

11. If two forces acting at a point be represented by the two diagonals of a parallelogram, their resultant will be represented by a line equal to twice one of the sides of the parallelogram.

12. If two forces be represented by the lines joining the bisections of two sides of a triangle with the opposite vertices, and both forces act either towards or from these vertices, then will the line joining the bisection of the third side and the opposite vertex represent the resultant.

13. ABC is an isosceles triangle, A and B the equal angles, CD a perpendicular from the vertex on the base; take $GD = \frac{1}{3}CD$; then if GA, GB represent two forces acting at a point G, GC will represent the force that will keep the point G at rest.

CHAPTER III.

THE TRIANGLE AND POLYGON OF FORCES

SECTION I.

Proof of the Triangle of Forces; and easy Exercises.

29. **The Triangle of Forces.**—*If three forces acting at a point be represented in magnitude and direction by the sides of a triangle taken in order, they will be in equilibrium.*

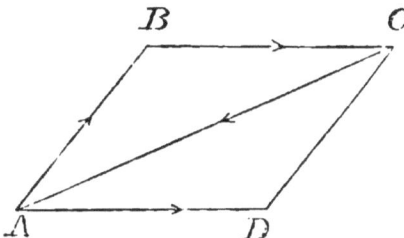

Let ABC be a triangle whose sides AB, BC, CA, taken in order, represent in magnitude and direction the forces AB, AD, CA, acting at the point A.

Complete the parallelogram ABCD.

Then the forces AB, AD, have a resultant AC, that is, the forces AB, AD, CA, are equivalent to AC, CA, and therefore balance each other.

It follows that the forces represented by AB, BC, CA, would be in equilibrium if they were applied directly at the point A.

30. **Conversely.**—*If three forces, acting at a point and keeping it at rest, be represented in direction by the sides of a triangle taken in order, these sides will be proportional to the magnitudes of the forces.*

This last statement governs the application of the proposition in practice. A triangle is constructed whose sides are parallel to the forces; the sides of this triangle are found, and give the relative magnitude of the forces.

31. Remarks on the preceding Proposition. The student will carefully bear in mind that the triangle ABC is not the body acted on by the forces, but that it is simply a triangle drawn on paper so that its sides represent the three forces *acting at the same point*. The sides of the triangle represent the forces in magnitude and direction, but not in line of action.* It will be shown hereafter that if three forces act in consecutive directions round a triangle, and be represented respectively by its sides, they cannot be in equilibrium.

EXAMPLES.

1. Can three forces represented in magnitude by 9, 6 and 4 pounds respectively, keep a particle at rest?

Since any two of the forces are greater than the third, three lines representing the forces would form a triangle (Euc. 1.22), therefore the three forces can keep a particle at rest.

2. Three forces represented by 1, 2, 3, act on a particle in directions parallel to the sides of a right-angled triangle taken in order, 1 and 2 being proportional to the sides, and 3 acting parallel to the hypothenuse. Will the particle remain at rest, and if not in what direction will it begin to move?

Since two of the forces are parallel and proportional to the sides of a right-angled triangle taken in order, if the particle is at rest the third force must be parallel and proportional to the hypothenuse, that is, the third force must be represented by $\sqrt{5}$; but the third force is represented by 3 which is greater than $\sqrt{5}$. The particle will therefore begin to move in the direction of the force 3.

---

*The line in which a force produces, or tends to produce motion, is called the line of the force's action; not only this line, but also every line that is parallel to this line, is said to be in the direction of the force.

EXERCISE I.

1. Enunciate and prove the triangle of forces.

Can a particle be kept at rest by three forces whose magnitudes are P, Q, and P + Q respectively?

2. Show that the following statement cannot always be true, "If three forces acting on a body are parallel to the sides of a triangle they will keep it at rest."

State the correct form of the theorem which this resembles.

3. Is it possible for three forces represented by the numbers 1, 4, 7, to a keep a particle at rest?

4. Show that when three forces are in equilibrium, no one of them is greater than the sum of the other two.

5. Three forces whose magnitudes are 6, 8, and 10 lbs., respectively, act upon a particle and keep it at rest; prove that the directions of two of the forces are at right angles to each other.

6. How can three sides of a triangle, not passing through a point, represent three forces which act at a point?

7. If a particle be kept at rest by three forces of 5 lbs., 6 lbs., and 7 lbs., respectively, draw lines that will correctly represent their directions.

8. Show how to keep a particle at rest by means of three forces, each equal to P lbs.

9. Two forces whose magnitudes are $\sqrt{3} \times P$, and P, respectively, act at a point in directions at right angles to each other; find the magnitude and direction of the force which will balance them.

10. If two forces, acting at right angles to each other, have a resultant, which is double the smaller of the two forces, find its direction.

11. If two forces be inclined to each other at an angle of 135°, find the ratio between them, when the resultant is equal to the smaller force.

SECTION II.

Extension of the Principle of the Triangle of Forces, and its application to find the tension of strings.

Ex. 3. A weight W is sustained by two cords of given lengths CA, CB, fastened to two points A and B, lying in the same horizontal line; to find the tensions in the cords, when they are at right angles.

Let $AC = a$, $BC = b$, then $AB = \sqrt{(a^2 + b^2)}$ (Euc. 1.47).

Let P, Q, be the tensions in BC and AC respectively.

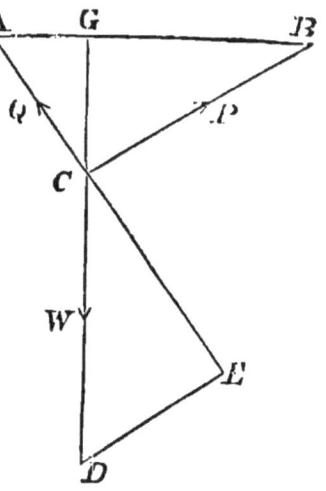

The point C is kept at rest by three forces, W acting vertically downwards, and P, Q, the tensions of the strings.

In the vertical line through C take CD, equal to AB, and through D draw DE, parallel to CB, meeting AC produced in E. Then in the triangles CDE, ABC, we have the side CD in the one equal to the side AB in the other, the right angle CED, equal to the right angle ACB, and the angle DCE, equal to the angle ABC, each being the complement of the angle BCG, therefore, the triangle CDE is equal in every respect to the triangle ABC (Euc. 1.26), and its sides are respectively parallel to the directions of the three forces W, P, Q, and therefore proportional to their magnitudes. Hence, by the Triangle of Forces,

$$\frac{P}{W} = \frac{DE}{CD} = \frac{AC}{AB} = \frac{a}{\sqrt{(a^2+b^2)}},$$

$$\therefore P = \frac{Wa}{\sqrt{(a^2+b^2)}}.$$

Similarly,
$$Q = \frac{Wb}{\sqrt{(a^2+b^2)}}.$$

32. Extension of the Principle of the Triangle of Forces. From the preceding article we see that
$$\frac{P}{W} = \frac{AC}{AB}, \quad \frac{Q}{W} = \frac{BC}{AB}, \text{ and } \frac{P}{Q} = \frac{AC}{BC},$$
or $\dfrac{P}{W} = \dfrac{\text{side perpendicular to direction of P}}{\text{side perpendicular to direction of W}}$;
and similarly for the others.

The above is only a particular case of a more general proposition. If the sides of a triangle be parallel and proportional to three forces which keep a particle at rest, it is evident that the triangle may be turned so that its sides shall be at right angles to their former position and still remain proportional to the forces. We may, therefore extend the enunciation of the triangle of forces given in art. 29, as follows :—

If three forces acting at a point be in equilibrium, and if any triangle be drawn, the sides of which are respectively parallel or perpendicular to their directions, the forces will be to one another as the sides of the triangle; and conversely, if the three forces are to one another as the sides of the triangle, they will be in equilibrium.

Ex. 4. If a string ACDB be 21 inches long; C and D two points in it, such that AC = 6, CD = 7; and if the extremities A and B be fastened to two points in the same horizontal line at a distance of 14 inches from each other; what must be the ratio of two weights, which, hung at C and D, will keep CD horizontal?

Let ACDB = 21,
AC = 6,
CD = 7,
DB = 8,
and AB = 14.
If AE = x,

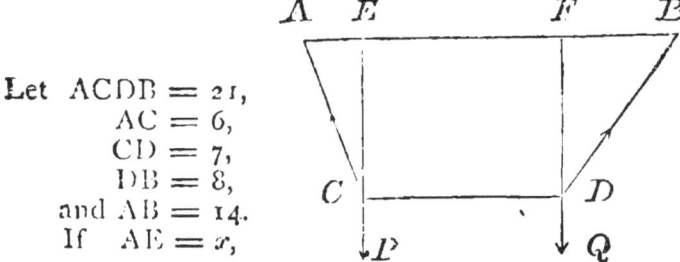

then $AC^2 - AE^2 = CE^2 = DF^2 = DB^2 - BF^2$;
or $6^2 - x^2 = 8^2 - \{14 - (x + 7)\}^2$
$= 8^2 - (7 - x)^2$;
$\therefore x = 1\frac{1}{2}$, and $BF = 7 - x = 5\frac{1}{2}$.

Let T be the tension of the string CD; the point C is kept at rest by three forces P, T, and the tension of AC, and these three forces are parallel to the sides of the triangle ACE, and are therefore proportional to them;

Hence, $\dfrac{P}{T} = \dfrac{CE}{AE}$.

Similarly, $\dfrac{T}{Q} = \dfrac{BF}{DF} = \dfrac{BF}{CE}$.

Multiplying, $\dfrac{P}{Q} = \dfrac{BF}{AE} = \dfrac{5\frac{1}{2}}{1\frac{1}{2}} = \dfrac{11}{3}$.

EXERCISE II.

1. A weight of 100lbs. is sustained by two cords whose lengths are 21 and 28 inches respectively, fastened to two points lying in the same horizontal line; required the tensions in the cords when they are at right angles to each other.

2. A weight of 10lbs. is suspended by two strings attached at their upper ends to two points in the same horizontal line; their lengths being 39 and 52 inches respectively, and the angle between them being 90°, find the tensions in each.

3. A weight of 24lbs. is suspended by two flexible strings, one of which is horizontal and the other is inclined at an angle of 30° to the vertical direction; what is the tension in each string?

4. The lower end B of a rigid rod without weight, 10ft. long is hinged to an upright post, and its other end A is fastened by a string 8ft. long to a point C vertically above B, so that ACB is a right angle. If a weight of 20cwt. be suspended from A, find the tension of the string.

5. A cord having equal weights attached to its extremities passes over pulleys A and B in the same hori-

zontal line 70 inches apart, and through a smooth ring C, from which a weight of 100lbs. is attached; what must be the magnitude of the equal weights, that C may rest exactly 12 inches below the line AB?

6. A thread 12 feet long is fastened at points A and B in the same horizontal line, 8ft. apart. At C and D points 4ft. and 5ft. respectively, from A and B, weights are attached; what must be the ratio of the weights that CD may be horizontal?

7. A and B are points 14 inches apart in a vertical wall, to which are attached the extremities of a cord ACDB. At C a weight of 11lbs. is attached, and at D another weight W. $AC = 6$ inches, $CD = 7$ inches, and $DB = 8$ inches; what must be the weight of W in order that CD may be horizontal?

SECTION III.

The reaction of hinges and smooth surfaces; and three forces keeping a body at rest.

33. **Reaction of Surfaces and Hinges.** It nearly always happens that amongst the forces which keep a body at rest is the reaction of one or more surfaces. Suppose a weight of 100lbs. to rest on a smooth table; the weight must be supported by the table which must, therefore, exert upwards a force of 100lbs. in a direction opposite to the direction of the weight. If we consider the case particularly, we shall see that this is an instance of what may be called a *distributive* force; for the under surface of the weight will be in contact with the table at many points, and at each point there will be a reaction. But it will be shown hereafter that in case of a force distributed over a surface, it is possible to find a single point and a single line such, that a certain force acting at that point in that line would produce the same effect as is really produced. This resultant reaction is called *the reaction of the surface.* The reaction of a smooth surface on a body in contact with it, is always exerted in a direc-

tion perpendicular to the surface. If two rods be hinged together the reaction at the hinge is generally unknown both as regards magnitude and direction ; but if a body is kept at rest by the reaction of a hinge and two other forces which are not parallel, the reaction of the hinge always passes through the intersection of the lines of action of the other two forces.

34. Three forces acting on a rigid body. *Whenever three forces which are not parallel act on a body and keep it at rest, their directions pass through the same point.*

Two, at least, of the forces meet in one point, and have a single resultant; this resultant must balance the third force. But two forces which balance must act in the same straight line and in opposite directions ; therefore the third force must pass through the intersection of the other two forces.

Ex. 5. A beam AB has one end attached to a hinge A, and the other end attached to a cord BC one end of which is tied to a peg. The weight of the beam is 50lbs. and may be supposed to act at its middle point. The beam and cord make angles of 60° on opposite sides of the vertical. Find the tension of the cord.

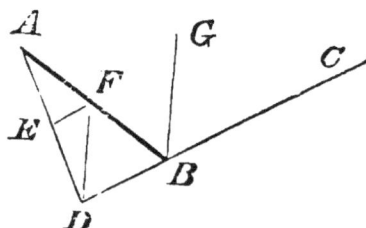

Let AB be the beam, F its middle point, and BG the vertical line.

Let T be the tension of the cord BC.

Through F draw a vertical line to represent the direction of the weight, meeting CB produced, in D.

The reaction of the hinge must pass through the point D, (art. 34).

Join AD, and through F draw EF parallel to BD.
The angle BFD = angle FBG = 60°, (Euc. 1.29).
And   „   FDB =   „   CBG, = 60°,   „
∴ triangle BFD is equilateral and
∴ DF = BF = AF,
∴ the angle FAD = angle FDA = ½ angle BFD = 30°;
∴ angle ADB = 90°.

Let DF = 2, then EF = 1, and ED = $\sqrt{3}$, (art. 28.)
The three sides of the triangle DEF are parallel to the three forces which keep the beam AB at rest, and are therefore proportional to them; hence
$$\frac{T}{50} = \frac{EF}{FD} = \tfrac{1}{2};$$
∴ T = 25 lbs.
Similarly the reaction of the hinge is found to be $25\sqrt{3}$.

EXERCISE III.

1. A sphere weighing 200 lbs. rests between two planes inclined to the horizon at angles of 30° and 60° respectively; find the pressures on the planes.

2. A carriage wheel whose weight is W and radius r, rests upon a level road; show that the force F necessary to draw the wheel over an obstacle of height h is
$$F = W \frac{\sqrt{(2rh - h^2)}}{r - h}$$

3. A rigid rod, the weight of which is 10 lbs., acting at its middle point, moves at one end about a hinge, and is supported at the other end by a piece of string attached to a point vertically over the hinge, and at a distance from it equal to the length of the rod; find the tension in the string when the rod rests in a horizontal position.

4. A rod AB without weight, can turn freely about a fixed point or hinge at one end B; it is held in a horizontal position by a force of 50 lbs. which acts vertically downwards through its middle point, and by a force P which acts at the end A, in such a manner that the angle BAP equals 30°; find P.

5. A rod AB whose weight is W, acting at its centre, and which can turn freely about a hinge at A, rests with its end against a smooth vertical wall, making an angle of 45° with it; find the direction and magnitude of the reaction at A.

6. A uniform beam AB, whose weight is W, and may be supposed to act at its middle point, rests with one end A against a smooth vertical wall; the other end B is supported by a string fastened to a point C in the wall. If the length of the beam be 3ft., and the length of the string 5ft.; find CA, and the tension of the string.

35. **The Polygon of Forces.** *If any number of forces acting on a particle be represented in magnitude and direction by the sides of a polygon, taken in order, they will be in equilibrium.*

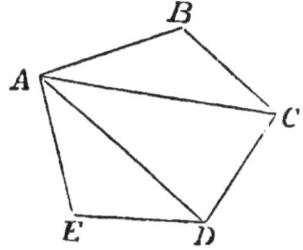

In the polygon ABCDE, from the Parallelogram of Forces, we know that the forces AB, BC are equivalent to a force represented by AC, that the forces AC, CD are equivalent to AD, and that AD, DE are equivalent to AE.

Therefore the forces represented by AB, BC, CD, DE, EA, are equivalent to AE, EA and will therefore balance each other. Hence the proposition is true.

The following converse of this proportion is also true, viz., *if any number of forces acting on a particle be in equilibrium, they can be represented in magnitude and direction by the sides of a polygon taken in order.*

36. **Resultant of any number of Forces acting at a point.** If any number of forces be in equilibrium, a force equal and opposite to any one will be the resultant of the remaining forces. Hence, any side of a polygon taken in reverse order, will represent the magnitude and direction of the resultant of any number of forces acting upon a point, when these forces are represented in magnitude and direction by the remaining sides of the polygon taken in order.

Thus if AB, BC, CD, and DE in the last figure represent in magnitude and direction, four forces acting upon a particle, the remaining side AE (not EA) will represent the magnitude and direction of the resultant.

37. Remarks on the Polygon of Forces. The remarks of art. 31 apply equally to the Polygon of Forces. If the forces were actually acting along the sides of the Polygon and represented by them in magnitude, they would clearly have a turning tendency, and could not produce equilibrium.

CHAPTER IV.

THE RESOLUTION OF FORCES.

SECTION I.

Method of resolving a given force into two others which act in given directions.

.38. Direct and Inverse problems. Direct problems are those in which the resultant of forces is to be found; inverse problems are those in which the components of a force are to be found. The former class is fixed and determinate; the latter is quite indefinite without limitations to be stated for each problem. A system of forces can produce only one effect; but an infinite number of systems can be obtained which will produce the same effect as one force. The problem, therefore, of finding components must be, in some way or other, limited. This may be done by giving the lines along which the components are to act.

Finding components is called *Resolution of Forces*.

39. *A force R acts on a particle A, it is required to resolve it into two others P and Q, which shall act in given directions.*

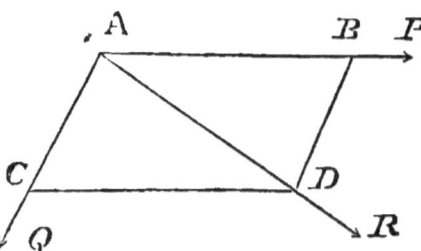

Let AR be the direction of the force R. Take AD representing R in magnitude. Through A draw AP,

AQ parallel to the given directions of P and Q respectively. Draw DB, DC parallel to AQ, AP respectively. Then AB, AC will evidently represent the required forces in magnitude and direction.

The following is a form of this proposition which is constantly occurring in statical problems.

40. **To find the effect of a Force in a given Direction.**

Let ABP, in the preceeding figure, be drawn from the point of application parallel to the required direction. Then we want to find what tendency the force R has to pull the particle along the line ABP.

From D draw DB perpendicular to ABP, and complete the paralellogram ABDC. Then the forces represented by AB, AC, are equivalent to the force R. But it is found by experiment that no force can produce any effect in a direction at right angles to itself. The force AC, therefore, being perpendicular to AB can produce no effect in the direction of AB. Therefore, the force AB represents the whole effect of R in a direction parallel to ABP.

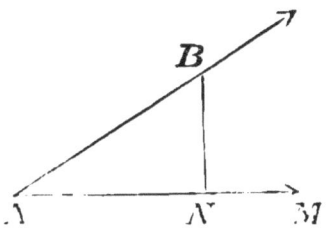

In all future investigations, if AB represents a force in magnitude and direction, and it is required to find the effect of that force along the line AM, draw BN perpendicular to AM, and AN will represent the portion of the force required.

41. **If the hypothenuse of a right-angled triangle represents a force, the sides, taken in order represent its components.** In any right-angled triangle ABN, if the hypothenuse AB, represents

a force in magnitude and direction, the sides AN, NB will represent its components. These components are called the *resolved parts* of AB in their respective directions.

EXAMPLES.

1. A man pulls a weight along a road by means of a rope, with a force of 20lbs. The rope makes an angle of 30° with the road; find the force he would need to apply parallel to the road to pull the weight.

In the fig. Art. 40, let AB represent the force of 20lbs. and let AM represent the direction of the road. From B draw BN perpendicular to AM. Then AN represents the required force. Since the angle BAN is 30°, and ANB is 90°, ABN will be 60°; and since AB represents a force of 20lbs. AN will represent a force of $10\sqrt{3}$ lbs. (art. 28), the force required.

2. Three forces act at a point, and include angles of 90° and 45°, respectively. The first two forces are each equal to 2P, and the resultant of them all is $P\sqrt{10}$; find the third force.

Let **Q** be third force.

Resolve Q vertically and horizontally, and we have for each component $\dfrac{Q}{\sqrt{2}}$.

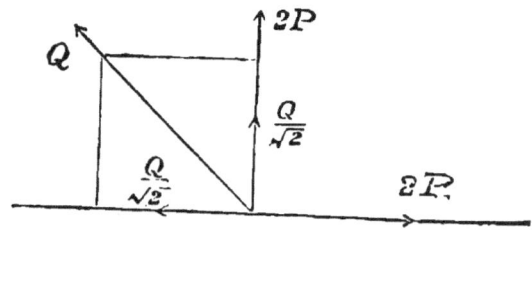

Then the sum of the vertical forces $= 2P + \dfrac{Q}{\sqrt{2}}$;

and algebraic sum of horizontal forces $= 2P - \dfrac{Q}{\sqrt{2}}$.

And the sum of the squares of these forces = square of resultant, that is

$$(2P + \frac{Q}{\sqrt{2}})^2 + (2P - \frac{Q}{\sqrt{2}})^2 = (P\sqrt{10})^2;$$
$$\therefore Q^2 = 2P^2,$$
or $Q = P\sqrt{2}$, the third force.

3. Show how it is possible for a sailing vessel to make way in a direction at right angles to that of the wind.

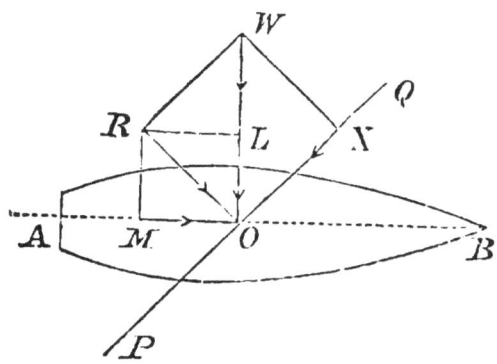

For simplicity let the sail be one of those attached to the yards of a ship, so that it extends on both sides of O.

It is evident that the sail must not be placed along the line AB, for then the only effect of the wind would be to blow the vessel sideways; nor could the sail be placed with its edge to the wind, that is, along the line OW, for then the wind would merely glide along the sail without producing any effect. Let, then, the sail be placed between the two positions, as in the direction PQ.

Let the line WO represent in magnitude and direction the force of the wind on the sail. Through O draw OR at right angles to PQ, and from W let fall the perpendiculars WX, WR, on PQ, OR respectively. Then the force WO can be resolved into the forces XO and RO. The force XO, in the direction of the plane of the sail, produces no effect in advancing the vessel, and may therefore be left out of consideration. The other component RO is perpendicular to the sail, but not in the direction in which the ship is sailing. Resolve RO into LO and MO. The effect of LO is to propel the vessel in a direction perpendicular to that in which she is sailing. This force is partly counteracted by the keel and the form in which the vessel is built, but more especially by the rudder which turns the head of the vessel towards the wind, and makes her sail sufficiently to the windward to counteract the effect of LO in driving her leeward. There remains MO which acts directly to push the vessel in the required direction. Hence it is seen how the wind, aided by the resistance of the water, is able to make the vessel move in a direction perpendicular to that in which the wind blows.

EXERCISE I.

1. Explain the meaning of the words Composition and Resolution of Forces, and show how forces may be compounded and resolved. A particle is acted on by a force whose magnitude is unknown, but whose direction makes an angle of $60°$ with the horizon. The horizontal component of the force is known to be 1.35; find the total force and also its vertical component.

2. Show how to resolve a given force into two components, one of which has a given magnitude and acts parallel to a given straight line.

As a special case, resolve a force of magnitude 12 acting horizontally from left to right, into two components one of which is a force of magnitude 25 acting vertically upwards.

3. Show why the traces of a horse ought, in general, to be parallel to the road along which he is pulling.

4. Show by a diagram the forces which keep a kite in equilibrium.

5. Show how it is possible for a sailing vessel to make way in a direction making half a right angle with that of the wind.

Why cannot a round tub be steered at as great an angle to the direction of the wind as a long-boat?

6. When a horse is employed to tow a barge along a canal, the tow-rope is usually of considerable length;

give a definite reason for using a long rope instead of a short one. Show whether the same considerations hold good in relation to the length of the rope when a steam-tug is used instead of a horse.

7. A horse walking by the side of a canal is to draw a boat along the canal by means of a horizontal rope attached to the boat near the bow. Point out the position in which the rudder must be placed in order that the boat may move parallel to the bank; and show by a diagram *all* the forces acting upon the boat when in motion.

8. Find the horizontal and vertical pressures, when a force of 100lbs. acts in a direction making an angle of 60° with the vertical line.

9. It is required to substitute for a given vertical force, two forces, one horizontal, the other inclined at an angle of 45° to the vertical; determine the magnitude of these two forces.

10. Find the horizontal and vertical pressures, when a force of 80lbs. acts in a direction making an angle of 30° with the vertical line.

11. Find the resultant of any three forces, the least of which is 10lbs., which are represented by, and act along OA, OB, OC, two sides and the diagonal of an oblong whose area is 60 square inches, and shorter side 5 inches.

12. Three forces of 20, 20, and 15lbs, respectively, act upon a particle; the angle between the first and second is 120°, and the angle between the second and third is 30°; find the magnitude of the resultant.

13. Three forces 99, 100, 101lbs. respectively act upon a particle in directions making an angle of 120° with each other, successively; find the magnitude of the resultant, and the angle it makes with the force 100.

14. A man and a boy pull a heavy weight by ropes inclined to the horizon at angles of 60° and 30° with forces of 80lbs. and 100lbs., respectively. The angle between the two vertical planes of the cords is 30°; find the single horizontal force that would produce the same effect.

SECTION II.

Conditions of equilibrium when any number of forces act at a point.

42. *To find the conditions of equilibrium when any number of forces act at a point, their directions being in one plane.*

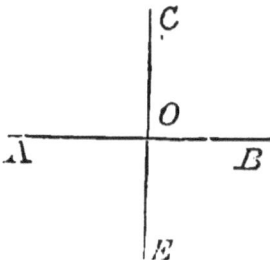

Let O be the particle on which the forces act. Through O draw any two lines AOB, and COE, at right angles to each other.

Resolve all the forces along AOB, and COE, respectively.

Let X_1 = sum of resolved pts. of fcs in the direction OB,
$\quad X_2 =\quad$ " $\quad$ " $\quad$ " $\quad$ OA,
$\quad Y_1 =\quad$ " $\quad$ " $\quad$ " $\quad$ OC,
$\quad Y_2 =\quad$ " $\quad$ " $\quad$ " $\quad$ OE.

Then in order that particle may be at rest, we must have separately,

$$X_1 = X_2,$$
$$Y_1 = Y_2.$$

Hence the conditions necessary and sufficient for equilibrium are as follows*:—

If any number of forces acting on a particle and maintaining equilibrium be resolved along any two lines at right angles to each other, the sums of the resolved parts in opposite directions along each of these lines, must separately be equal.

---

* By the words "necessary and sufficient for equilibrium" is meant that, on the one hand, if the forces are in equilibrium the above equations will be satisfied, and on the other hand, if the above equations are satisfied the forces will be in equilibrium.

RESOLUTION OF FORCES.

Since $X_1 - X_2 = 0,$
and $Y_1 - Y_2 = 0,$
then, if we consider forces acting in one direction *positive* and those acting in the opposite direction *negative*, the conditions of equilibrium may be stated as follows:—

The algebraic sums of the forces resolved into two perpendicular directions shall separately vanish.

EXAMPLES.

1. A weight of 10lbs. is supported by two strings each of which is 3 ft. long, the ends being attached to two points in a horizontal line 3 ft. apart; find the tension of each string.

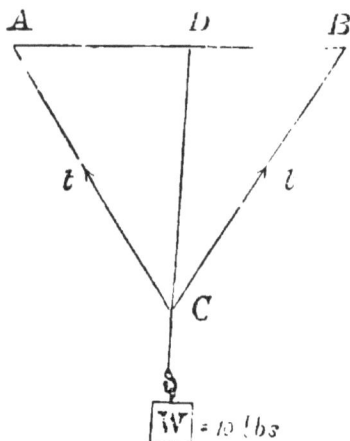

Let A and B be the points of support
Let AC and BC be the strings.
Let W be the weight of the body.
Draw CD perpendicular to AB. ABC is an equilateral triangle, and therefore,

the angle CAD = angle CBD = 60°,
and angle ACD = angle BCD = 30°.
The tension in AC = tension in BC = t.

Resolve each of the tensions along CD. To do this, suppose CB = 2, then CD = $\sqrt{3}$; if, therefore, the force acting along CB were 2, its resolved part along CD would be $\sqrt{3}$ (art. 28); but the force acting along CB is t, its resolved part along CD is therefore $\frac{1}{2}t\sqrt{3}$. Since there is equilibrium, the sum of the resolved parts must be equal to the weight, (art. 42); hence

$$2\left(\frac{\sqrt{3}}{2}\right) = 10,$$

$$\therefore t = \frac{10}{\sqrt{3}} \text{ lbs.}$$

43. Resolution along one line often sufficient.
The student will observe that in the solution of problems in which the forces are in equilibrium it is frequently sufficient to resolve the forces along one line only. Any line whatever may be chosen and then the sums of the resolved parts, in opposite directions, along this line must be equal to each other. The line must be chosen so as to make the resulting equation as simple as possible. In the preceding article we inferred from the symmetry of the figure that the tension of each string was the same, and therefore the vertical resolution was sufficient. Had the angles ACD, BCD been unequal, it would have been necessary to have resolved *vertically*, and *horizontally*, and from the two equations thus obtained, the two tensions could have been determined.

2. A weight W is sustained on an inclined plane by a certain force; the inclination of the force to the inclined plane is 30°, and the inclination of the plane to the horizon is 30°: find the force and the pressure on the plane.

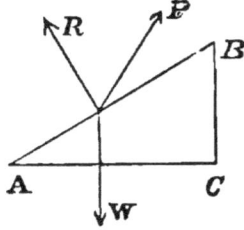
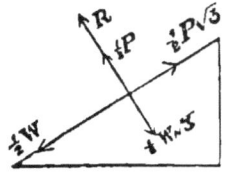

Let ABC be the inclined plane.
" R = the reaction of the plane.
" P = the sustaining force.

Then, resolving the forces along and perpendicularly to the inclined plane, we have

$$P \frac{\sqrt{3}}{2} = \tfrac{1}{2}W \ldots \ldots (1),$$

and $\qquad R + \tfrac{1}{2}P = W\frac{\sqrt{3}}{2} \ldots . (2).$

We have from (1), $P = \dfrac{W}{\sqrt{3}}.$

Substitute in (2), and $R = \dfrac{W}{\sqrt{3}}.$

The second figure shows the two groups into which by resolution we have decomposed the three forces P, W, R; the equations (1) and (2) representing the conditions for equilibruim of the two new groups, each regarded as independent of the other.

EXERCISE II.

1. What power acting parallel to a smooth plane inclined to the horizon at an angle of $30°$, will sustain a weight of 4lbs. on the plane?

2. What force acting horizontally, will sustain a weight of 12lbs. on a plane inclined to the horizon at an angle of $60°$?

3. A cord is attached to two fixed points A and B in the same horizontal line, and bears a ring, weighing 10lbs., at C, so that ACB is a right angle; find the tension of the cord.

4. Two strings at right angles to each other support a weight, and one string makes an angle of $30°$ with the vertical line. Compare the tensions of the strings.

5. A weight of 24lbs. is suspended by two flexible strings, one of which is horizontal and the other inclined at an angle of $30°$ to the vertical direction; find the tension of each string.

6. Two pictures of equal weights are suspended symmetrically by cords passing over smooth pegs; the two portions of the cord in one case making an angle of $60°$, and in the other an angle of $120°$ with each other. Compare the tensions of the strings.

7. A force of 40lbs. acting on a particle between two forces of 20lbs. and $20\sqrt{3}$lbs. respectively, makes an angle of 60° with the former and 30° with the latter; find the magnitude and direction of the force which will keep the particle at rest.

8. On AB an inclined plane whose base is AC and which has the angle BAC equal to $\frac{2}{3}$ of a right angle, a heavy body is kept at rest by two equal forces, the one acting in the direction of AB, and the other in the direction of AC and towards the plane. Prove that the reaction of the plane on the body is equal to the weight of the body.

9. A weight W is supported on an inclined plane, the inclination of which to the horizon is 60°, by three forces each equal to P, one acting along the plane, another horizontally, and a third in a direction inclined to the plane at an angle equal to the inclination of the plane; required the value of P, and the pressure on the plane.

10. Two strings fastened to pegs A and B in a vertical wall, of which A is the higher, are joined at the point C to a weight W, the strings are of such lengths that BC is horizontal, and the angle BCA is 135°; find the tensions in the strings.

11. Two planes of equal altitude are inclined at angles of 60° and 45° to the horizon; what weight resting on the latter will balance 20lbs. on the former, the weights being connected by means of a string passing over the common vertex?

CHAPTER V.

PARALLEL FORCES.

44. Definition of Parallel Forces. Parallel forces are those acting at different points of a body and in directions parallel to one another.

45. *To find the resultant of two parallel forces acting on a rigid body in the same direction.*

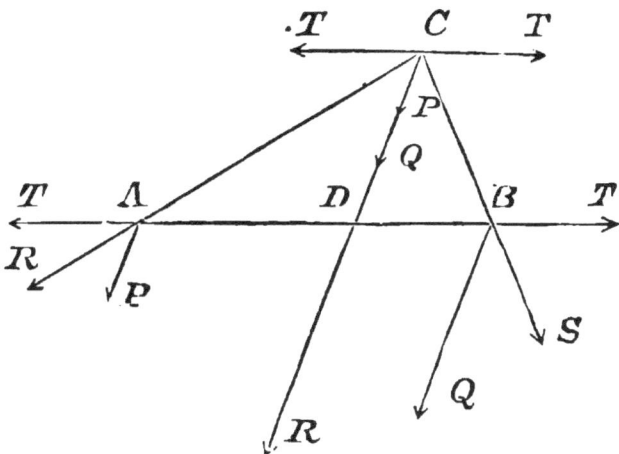

1. **The magnitude of the resultant.** Let P and Q represent two parallel forces acting on a rigid body, at points A and B and in the same direction.

Join AB; and apply at A and B two forces, T, T, acting in opposite directions in the line AB; this will not affect the equilibrium of the system.

Let the forces T and P at A have a resultant R, and the forces T and Q at B have a resultant S. Let the forces R and S meet in C. They can now be here resolved into their original components P, T, and Q, T.

The forces T, T, acting at C will balance and may be removed. There will remain P + Q acting at C, in a direction CD parallel to AP or BQ and this is the resultant of P and Q acting at A and B repectively.

Let R represent the resultant of P and Q, then

$$R = P + Q.$$

Hence, *the resultant of two parallel forces P and Q, acting in the same direction, is equal to their sum and acts parallel to their directions in a straight line which cuts AB in D; so that it may be supposed to act at D.*

2. **The position of the point D.** The sides of the triangle ACD are parallel to the three forces P, T, R, which keep the point A at rest, and are, therefore, proportional to them; hence by Art. 30 we have

$$\frac{P}{T} = \frac{CD}{AD}\ldots\ldots(1).$$

Similarly
$$\frac{T}{Q} = \frac{BD}{CD}\ldots\ldots(2).$$

Multiplying (1) and (2) we have

$$\frac{P}{Q} = \frac{BD}{AD};$$

that is, *the point D divides AB into segements which are inversely as the forces P and Q respectively.*

Let AB = a, and AD = x.

Since
$$\frac{P}{Q} = \frac{BD}{AD},$$

or P.AD = Q.BD;

that is P.x = Q ($a - x$);

therefore
$$x = \frac{Qa}{P+Q}.$$

46. *To find the resultant of two parallel forces acting on a rigid body and in contrary directions.**

---

* *Contrary* signifies parallel and in dissimilar directions; *opposite* signifies, contrary and in one line.

Let P and Q be two parallel forces acting in contrary directions, and let Q be greater than P.

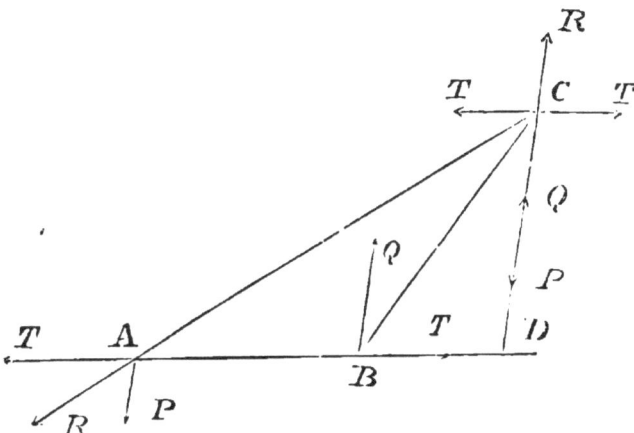

By the same method as in the preceding article it may be shown that
$$R = Q - P,$$
and that it acts parallel to the directions of P and Q in a straight line which cuts AB produced at a point D, such that
$$\frac{P}{Q} = \frac{BD}{AD};$$
that is, *D divides AB produced through B into segments which are inversely as the forces P and Q respectively.**

Let $AB = a$, and $AD = x$.

Since
$$\frac{P}{Q} = \frac{BD}{AD},$$
or
$$P.AD = Q.BD;$$
that is
$$P x = Q(x - a);$$
therefore
$$x = \frac{Qa}{Q - P}.$$

---

*When a line is cut at any point the intercepts between the point of section and its extremities are called its segments. When the point of section lies between the extremities of the line it is said to be cut *internally*; but when it is not the line itself but its production that is cut, and therefore the point of section lies beyond one of its extremities, it is said to be cut *externally*.

47. Resultant of two parallel forces. If we extend to parallel forces the same method of indicating opposition of direction by difference of sign, which was used in the case of forces acting in the same line (art. 42), we can include the results of the two preceding articles in the following statement:—

The resultant of two parallel forces is equal to their algebraic sum, and is in the parallel line which divides any line drawn across their lines of action into segments inversely as their magnitudes.

48. Resultant of any number of parallel forces. We may find the resultant of any number of parallel forces by repeated application of the process of the last two articles. First find the resultant of two of the forces; and then find the resultant of this and the third force; and so on.

49. Couples. In Article 46 it was shown that if P and Q are two parallel forces acting at A and B, and in contrary directions, then if Q is greater than P, their resultant R will be a parallel force acting in the same direction as P and Q through a point D, given by the equation,

$$AD = \frac{Qa}{Q-P},$$

Now, if we suppose P to be gradually increased, but a and Q to remain unaltered, the magnitude of R, (or $Q-P$), will continually diminish and AD will continually increase and when P becomes equal to Q, the point D is removed to an infinite distance, and $R = 0$; yet the forces are not in equilibrium, since they are not directly opposed. Hence two equal and contrary forces neither balance nor have a single resultant. It is clear that they have a tendency to turn the body to which they are applied. Such a pair of forces constitute what is called a *couple*. This case must be excluded from the statement in Art. 47.

50 Centre of Parallel Forces. The expressions for x or AD in Articles 45 and 46 are independent of the angle which the direction of the forces makes with the line joining their points of application. The position of the point D is not, therefore, altered by changing the direction of the forces, if the points of application remain the same. Generally, if we conceive any system of Parallel Forces, and suppose that each force acts at a particular point, then if we suppose the lines along which the forces act to be turned around the points through any equal angles so that they still continue parallel, it will be found that there is a certain fixed point through which their resultant will always pass, whatever be the magnitude of the angles; this fixed point in the line of action of the resultant is called *the centre of that system of parallel forces*. If the parallel forces are the weights of the parts of a heavy body, or of the members of a system of heavy bodies, the centre of these parallel forces is the centre of gravity of the body or system of bodies.

EXAMPLES.

1. Two persons A and B, carry a weight of 100lbs. on a pole between them. The weight being placed two feet from A and three feet from B, find what portion of it they respectively support.

Let C be the point of the rod at which the weight W is hung, and let R, R' be the forces which A and B respectively exert at the ends of the pole in a vertical direction; by supposition, these with W acting downwards at C preserve

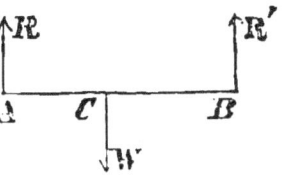

equilibrium in the system; therefore W is equal and opposite to the resultant of R and R'; or

$$W = R + R';$$

and $$\frac{R}{R'} = \frac{BC}{AC} = \frac{3}{2}.$$

Hence we obtain $R = \frac{3}{5} W = 60$lbs.;

and $$R' = \frac{2}{5} W = 40\text{lbs}.$$

2. The resultant of two parallel forces acting in contrary directions is 12lbs., and acts at a distance of 5in. from the greater and 7in. from the lesser force, find the forces.

Let P = the greater force,

then $R = P - Q = 12$,
or $P = 12 + Q$,
And $5P = 7Q$,
or $5(12 + Q) = 7Q$;
therefore $Q = 30$lbs.;
and $P = 42$lbs.

EXERCISE I.

1. Two men of the same height, carry on their shoulders a pole 6ft. long, and a weight of 121lbs. is slung on it, 30in. from one of the men; what portion of the weight does each man support?

2. Two men carry a weight of 152lbs. between them on a pole, resting on a shoulder of each; the weight is three times as far from one as from the other; find how much weight each supports, the weight of the pole being disregarded.

3. Two men are carrying a block of iron, weighing 176lbs. suspended from a uniform pole 14ft. long; each man's shoulder is 1ft. 6in. from his end of the pole. At what point of the pole must the heavy weight be suspended, in order that one of the men may bear $\frac{4}{5}$ of the weight borne by the other?

4. The resultant of two parallel forces acting in contrary directions is 6lbs., acting 8in. from the greater force, which is 10lbs.; find the distance between them.

5. A weight of 12lbs. is placed on a square board 5in. from one side and 10in. from the opposite side, along the line that bisects its width. The board is suspended at its corners by four vertical strings; find the tensions in each string.

6. A flat board 12in. square, whose weight is 12lbs., acting at its centre, is suspended in a horizontal position by strings attached to its four corners A, B, C, D, and a weight equal to the weight of the board is laid upon it at a point 3in. distant from the side AB, and 4in. from AD; find the tensions in the four strings.

7. A weight of 15olbs. rests on a triangular table at a point in the line joining one of the angular points with the middle of the opposite side, and at a distance from the angular point equal to twice its distance from the side. The table is supported on three props at its angular points ; find the pressure on each prop.

8. A beam AB, 1oft. long, rests horizontally upon two vertical props, A and B, and another beam CD, 20ft. in length, rests upon two vertical props, C and D; and a third beam 3oft. in length, lies across the two former beams in such a way that the one extremity E is 3ft. from the prop A, and the other extremity F, is 5 feet from the prop D; a weight of 6olbs. is attached to the third beam at a distance of 1oft. from F; find the pressure on A, B, C, D respectively.

9. If a weight rests in the middle of a square rough table, will the pressure on each leg be altered if one pair of legs are longer than the opposite pair?

10. Two forces of 1olbs. each, act on a body along parallel lines, and in contrary directions ; why should it be impossible to balance these forces by any one force ?

11. The weight of a window-sash 3ft. wide is 5lbs., acting at the middle point of the sash ; each of the weights attached to the cord is 2lbs. ; if one of the cords be broken, find at what distance from the middle of the sash the hand must be placed to raise it with the least effort.

CHAPTER VI.

MOMENTS OF FORCES.

SECTION I.

Principle of moments. Equilibrium of a body capable of turning round a fixed point. Equilibrium of a body acted on by any number of forces in one plane. Hints for the solution of the problems.

51. **Introduction.** The moment of a force about a given point is its tendency to produce rotation about the point.

Suppose a rod OD capable of turning about the fixed point O, to be acted on by a force P; it is shown by experiment that the tendency of the force to turn the rod about O depends on the magnitude of the force and on its distance from the point O. We might, for example, double this tendency either by doubling the force, or by keeping the force the same and causing it to act at twice the distance from O. Hence the tendency of the force to turn the rod about O is measured by the product of the force into the perpendicular OD.

When one point in a body is fixed, in order that the body may be at rest it is evident that the moment in one direction about that point must be equal to the moment in the opposite direction.

But when a body is at rest under the action of forces it is evident that we may imagine any point in it to be fixed, for the fixing of a point which is already at rest without introducing any new forces at the point would not disturb the equilibrium. Hence, the tendency to turn in one direction about any point must be equal to that in the opposite direction, or if we consider the moments in one direction *positive*, and those in the opposite direction *negative*, then the algebraic sum of the moments of the forces about *any* point must be zero. This is one of the most important principles in statics. It can be rigidly deduced from the parallelogram of forces. Its demonstration and applications form the subject of the present chapter.

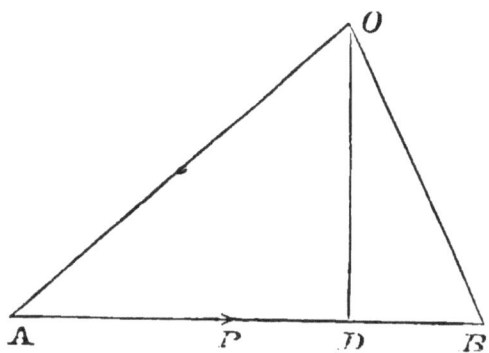

52. Definition of the moment of a force. Let AB represent any force P and let O be any point in the same plane. From O let fall a perpendicular OD on AB. Then AB.OD or P.OD is called the moment of P about the point O. Hence we have the following definition:

The moment of a force about a given point is the product of the force into the length of the perpendicular drawn from the given point upon the direction of the force.

53. Geometrical measure of the moment of a force. Since the area of any triangle is equal to half the product of the base into the altitude, the moment of P with respect to O is represented by twice the area of the triangle of AOB. Hence, *the moment of any force may be represented by twice the area of the triangle having for its base the straight line which represents the force, and for its vertex the point about which the moment is taken.*

54. **Positive and negative moments.** If the moment of a force tending to turn a body round a fixed point in one direction be considered *positive*, then the moment of a force tending to turn the body in an opposite direction will be considered *negative*. It is indifferent in any investigation which kind of moment we consider positive and which negative; but when a choice has been made we must keep to it during that investigation.

55. *The moment of the resultant of two intersecting forces round any point in their plane is equal to the algebraic sum of the moments of the forces about the same point.*

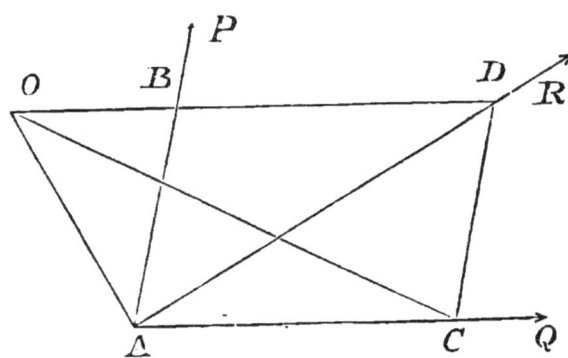

Let P and Q be two forces whose directions intersect in A.

1. Let the point O about which the moments are to be taken fall *without* the angle PAR, as shown in the annexed figure; in this case all the moments are positive; we have therefore to show that

moment of R = moment of P + moment of Q.

Draw OB parallel to AQ. Divide AB into as many equal parts as there are units in the force P, and take

AC as many of these equal parts as there are units in Q; then AB and AC will represent the forces P and Q.

Complete the parallelogram of which AB and AC are adjacent sides and join AD; then AD will represent the resultant of P and Q.

Join OA and OC. Then we have
moment of R = 2 triangle AOD
= 2 AOB + 2 ABD
= 2 AOB + 2 ADC
= 2 AOB + 2 AOC
= moment of P + moment of Q.

2. Let the point O fall within the angle PAR as shown in the annexed figure; in this case the moments of Q and R with respect to O are positive, and that of P negative, so that we have to prove that

moment of R = — moment of P + moment of Q.

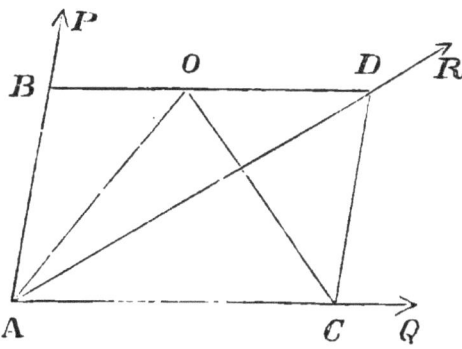

As before draw OB parallel to AQ.

Divide AB into as many equal parts as there are units in the force P; and take AC as many of these parts as there are units in Q; then AB and AC will represent the forces P and Q.

Complete the parallelogram of which AB and AC are adjacent sides, and join AD; then AD will represent the resultant of P and Q.

Join OA and OC. Then we have

moment of R = 2 triangle AOD
= 2 ABD — 2 AOB
= 2 ADC — 2 AOB
= 2 AOC — 2 AOB
= moment of Q — moment of P
= algebraic sum of moments of P and Q.

Similiar results may be obtained for other positions of the point O, and the student will find it instructive to work out the different cases. Hence the proposition is true.

56. Moments about a point in the direction of the resultant. Let O be any point in the direction of R, then the moment of R round this point vanishes, and we have the following important proposition:—

If any point be taken in the direction of the resultant of two intersecting forces, the moments of the forces about this point will be equal and opposite.

57 *The moment of the resultant of any two parallel forces about any point in their plane, is equal to the algebraic sum of the moments of the forces about the same point.*

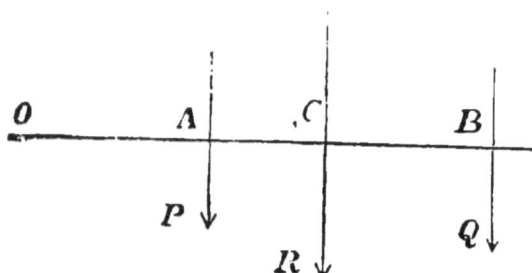

Let P and Q be two forces acting along parallel lines, and let O be any point in their plane.

From O draw OAB perpendicular to the directions of P and Q respectively.

If R be the resultant of P and Q, then
$$OC.R = OC (P + Q), \text{ since } R = P + Q.$$
$$= OC.P + OC.Q$$
$$= (OA + AC) P + (OB - CB) Q$$
$$= OA.P + OB.Q + AC.P - CB.Q$$
but $AC.P - CB.Q = 0$ (art. 45);
$$\therefore OC.R = OA.P + OB.Q.$$

Now OC.R, OA.P, OB.Q, are the moments, with respect to O of R, P, Q respectively, and the above equation shows that the moment of R is equal to the sum of the moments P and Q. A similar proof will apply to every position of O, and to cases in which P and Q act in contrary directions.

58. **Extension of the preceding propositions to any number of Forces.** The preceding propositions hold for *any* number of forces in one plane. For since the sum of the moments of the two forces is equal to the moment of their resultant, we may substitute the resultant for the two forces; we can now combine this resultant with a third force, and so on for any number of forces. Hence, we have the following important proposition :—

Principle of moments. **The moment of the resultant of any number of forces about a given point, in one plane, is equal to the algebraic sum of the moments of the forces with respect to the same point.**

59. *Equilibrium of a rigid body capable of turning round a fixed point or axis.*

(1.) When there is equilibrium the resultant of all the forces must pass through the fixed point; and conversely when the resultant passes through the fixed point there will be equilibrium.

If the resultant passes through a fixed point, its moment about that point is zero, hence when there is equilibrium,

(2.) The algebraic sum of the moments of the forces about the fixed point is equal to zero.

The last proposition is of great importance. It is more conveniently applied to the solution of mechanical problems when stated as follows :—

If any number of forces acting in the same plane keep a body in equilibrium round a fixed point, and if their moments with reference to that point be taken, the sum of the moments of those forces which tend to turn the body from right to left round the fixed point, will equal the sum of the moments of those forces which tend to turn the body from left to right.

60. **Any point in a body at rest may be considered a fixed point.** In applying the above proposition, the student will bear in mind the remark made in the introduction to this chapter, viz., that when a body is at rest under the action of any forces we may imagine any point in it, or rigidly connected with it, to be fixed; then this point will be the "*fixed point.*"

61. *Equilibrium of a rigid body acted on by any number of forces in one plane.*

If the forces be resolved into two directions at right angles to each other, then in order that there may be equilibrium, the algebraic sums of the forces in these directions must separately vanish. But this is not sufficient; for the algebraic sum of a number of forces would vanish if the sum of the forces in one direction were equal to the sum of the forces in the contrary direction. Now instead of these forces we may substitute their resultants. We have then two equal forces acting in opposite directions which will not produce equilibrium unless they act at the same point. If they do not act at the same point they form a couple and produce rotation (art. 49). Now in order that there should be no rotation it is necessary that the sum of the moments of the forces which tend to turn the body in one direction, round any point in their plane, shall be equal to the sum of the moments of the forces which tend to turn the body in the opposite direction round the same point. Hence the conditions necessary and sufficient for equilibrium when any number of forces act on a rigid body in one plane are as follows:—

(1.) The algebraic sums of the forces resolved into two directions at right angles to each other must separately vanish, and

(2.) The algebraic sum of the moments of the forces about any point in their plane must also vanish.

62. **Hints for the Solution of Problems.** In applying the principles of this chapter to the solution of problems, the following summary of facts and hints may be found useful.

I. When two forces are in equilibrium they must act in the same straight line and in opposite directions (art. 22).

II. When three forces, not parallel, are in equilibrium, their lines of action must meet in a point (art. 34).

III. Reaction of smooth surfaces are perpendicular to the surfaces. Reactions of hinges are generally unknown (art. 33). If a rod rests on a smooth peg the reaction of the peg is at right angles to the rod.

IV. Read the problem carefully. Draw a figure of the system of forces, which keep the body at rest, as accurately as you can, representing all the forces by straight lines and their directions by arrows.

V. Resolve the forces acting upon the body along two lines at right angles to each other. Then, the sums of the resolved parts in opposite directions along each of these lines will separately be equal. This will give you two equations. Take moments about some point. Put the sum of the moments, which tend to turn the body in one direction about the point, equal to the sum of the moments which tend to turn the body, in the opposite direction, about the same point. This will give a third equation. If the number of unknown quantities in these equations exceed the number of the equations, additional equations must be obtained from the geometrical relations of the figure.

VI. In resolving forces, and taking moments, equations are often much simplified by choosing the directions of resolution, and the points about which moments are to be taken, so that forces not required may vanish.

1. Forces should be resolved at right angles to unknown reactions.

2. Moments should be taken about points in directions common to as many forces as possible.

VII. The general methods here indicated may be often abbreviated by particular artifices peculiar to each problem. The following may be noticed:—

1. When there are two unknown forces, an equation may be found containing only one, either by taking moments about some point in the other, or by resolving the forces in a direction at right angles to this other.

2. When there are three unknown forces, one which is not required may be excluded, for two equations may be found by resolving in a direction perpendicular to that of the unknown force, and by taking moments about a point in the line of its direction.

VIII. The student will derive much benefit by solving a problem in various ways. Thus, he may assume new directions of resolution, and a new point about which to take moments; or he may endeavor to abbreviate the solution by a geometrical construction. One problem thoroughly understood will give a clearer insight into the principles of Statics than the imperfect comprehension of many.

EXERCISE I.

1. What is meant by the moment of a force about a given point? How is its magnitude determined?

2. Show how the moment of a force with respect to a point may be represented by an area.

3. In the case of two forces which act along intersecting lines, show that the sum of the moments of the forces with respect to any point in the plane of the lines equals the moment of their resultant with respect to the same point taking the case in which all the moments are not positive.

What assumption must be made respecting the signs of the moments, in order that the above statement may include all cases?

4. State the Principle of Moments, and show by it that if three forces act on a body, and the directions of two of them pass through a given point, there cannot be equilibrium unless the direction of the third force passes through the fixed point.

5. A man carries a bundle at the end of a stick over his shoulder, and the portion of the stick between his shoulder and his hand is shortened, show that the pressure on his shoulder is increased. Does this change alter his pressure on the ground?

6. If a man wants to help a waggon up a hill, is there any mechanical reason why he should put his shoulder to the wheel instead of pushing at the body of the waggon? and if so, show at what part of the wheel force can be applied with the greatest effect.

7. Three parallel forces keep a rigid body at rest; find the relation subsisting between their magnitudes, directions, and distances.

8. When three forces, acting in one plane, on a rigid body, produce equilibrium, the algebraic sum of the moments of either pair about any point in the line of action of the third is zero.

9. Find the conditions of equilibrium of any system of forces acting in one plane. Account for one of these conditions only being necessary in the case of a body capable of turning about a fixed point.

SECTION II.

The application of the Principle of Moments to the Lever. Examples and Exercises.

63. **The Lever.** This is the name given to a rigid rod capable of turning round a fixed point called the *fulcrum*. The parts into which the rod is divided by the fulcrum are called the *arms* of the lever. When the arms are in a straight line, the rod is called a *straight* lever; when the arms are not in a straight line it is called a *bent* lever.

64. **Recapitulation.** If two forces act upon a lever, supposed to be without weight, and produce equilibrium,

their resultant must pass through the fulcrum; and conversely, if their resultant pass through the fulcrum, there will be equilibrium.

The above condition is equivalent to the following one:—

The moments of the forces about the fulcrum shall be equal and opposite.

If several forces act on the lever and produce equilibrium, the resultant of all the forces must pass through the fulcrum; and conversely, if their resultant pass through the fulcrum, there will be equilibrium.

The above condition may be replaced by the following one:—

The sum of the moments of all the forces which tend to turn the lever in one direction about the fulcrum shall be equal to the sum of the moments of all the forces which tend to turn the lever about the fulcrum in the opposite direction.

EXAMPLES.

1. Four weights of 2lbs., 6lbs., 14lbs. and 10lbs., are placed at equal distances on a straight lever; find the position of the fulcrum when the lever is 21 inches long and the weights 2lbs. and 10lbs. are placed at its extremities; supposing the lever to be without weight.

```
   2        6       1·4      10
―――――――――――――――――――――――――――――――
   A        B        C       D
```

Let AD be the lever, and let the weights act at A, B, C, D respectively.

Let $x =$ the distance from A at which the resultant acts.

Take moments about A; then since the resultant equals the sum of the forces, and the moment of the resultant about any point equals the sum of the moments of the forces about the same point, we have

$32x = 2 \times 0 + 6 \times 7 + 14 \times 14 + 10 \times 21$
$= 448$;
$\therefore x = 14$ inches.

The fulcrum, therefore, coincides with the point C.

2. A bent lever without weight consists of two arms, one of which is twice as long as the other, and inclined to each other at an angle of 120°; find the ratio of the weights that must be suspended from their ends, so that the lever may rest with the shorter arm horizontal.

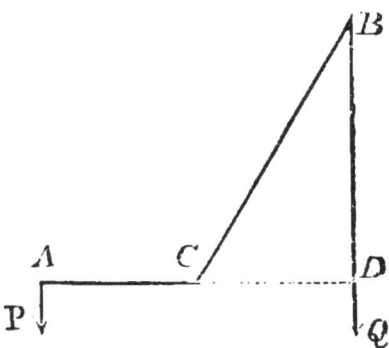

Let ACB be the lever, having $AC = a$, and $CB = 2a$.
The angle $ACB = 120°$, and $\therefore BCD = 60°$, and $CBD = 30°$.
Then CD will $= a$ (Art. 28).
Take moments about C,
and   P.AC = Q.CD.
that is   P.a = Q.a;
therefore   P = Q.

3. A heavy uniform beam, the weight of which may be considered as a force acting at its middle point, rests upon the shoulders of two men, one of whom stands at the end of the beam. Where must the other stand so that he may bear twice as much as the first?

Let AB be the beam, and $a =$ its length.

Let W = the weight of the beam, acting at its middle point C.

Suppose the first man to stand at A, and the weight supported by him to be P, and suppose the second man to stand at D, the weight supported by him will be 2P.

Then $\quad P + 2P = W;$

or $\quad\quad P = \dfrac{W}{3}.$

Take moments about A,

and $\quad W \times \dfrac{a}{2} = 2P \times AD$

$\quad\quad\quad\quad = \dfrac{2W}{3} \times AD;$

therefore $\quad AD = \dfrac{3a}{4}.$

Or thus: take moments about C,

and $\quad P \times \dfrac{a}{2} = 2P \times CD;$

therefore $\quad CD = \dfrac{a}{4};$

and $\quad AD = AC + CD$

$\quad\quad\quad = \dfrac{a}{2} + \dfrac{a}{4}$

$\quad\quad\quad = \dfrac{3a}{4}.$

EXERCISE II.

1. A lever is held in a horizontal position by two supports that are 5ft. apart, and a weight of 10lbs. is hung at the distance of $3\frac{1}{2}$ft. from one of the supports; find the pressure sustained by the other.

2. A lever 7ft. long is supported in a horizontal position by props placed at its extremities. Where must a weight of 56lbs. be placed on it so that the pressure on one of the props may be 8lbs. ?

3. On a horizontal rod, 45in. long, whose extremities are supported, where must a weight be placed so that the pressure on the supports may be as 5 to 4?

4. Two men of the same height bear a weight suspended from a pole, which rests on their shoulders. Where must the weight be placed so that one man may support twice as much as the other?

5. If the forces at the end of the arms of a horizontal lever be 8lbs. and 7lbs. and the arms 8in. and 9in. respectively; find at what point a force of 1lb. must be applied perpendicularly to the lever to keep it at rest.

6. A lever with a fulcrum at one end is 3ft. in length. A weight of 28lbs. is suspended from the other end. If the weight of the lever is 2lbs. and acts at its middle point, at what distance from the fulcrum will an upwards force of 5clbs. preserve equilibrium?

7. The length of a horizontal lever is 12ft. and the balancing weights at its ends are 3lbs. and 6lbs. respectively; if each weight be moved 2ft. from the end of the lever, find how far the fulcrum must be moved for equilibrium.

8. The whole length of each oar of a boat is 10ft., and from the hand to the rowlock the distance is 2ft. 6in.; each of 8 men sitting in the boat pulls his oar with a force of 50lbs. Supposing the blades of the oar not to move through the water, find the resultant force propelling the boat.

9. A straight rod, movable about one end, makes an angle of 30° with the vertical. A weight of 7lbs. hangs at the other end, what force acting perpendicularly to the rod at its middle point will preserve equilibrium?

10. A straight lever is inclined at an angle of 60° to the horizon and a weight of 360lbs. hung freely at the distance of 2in. from the fulcrum is supported by a power acting at an angle of 60° with the lever, at the distance of 2ft. on the other side of the fulcrum; find the power.

11. ACB is a bent lever; the arms CA, CB, are straight and equal, and inclined to one another at an angle of 135°. When CA is horizontal a weight of P

at A just sustains a weight of W at B; and when CB is horizontal the weight W at B requires a weight Q at A to balance it; find the ratio of Q to P.

12. A bent lever has equal arms making an angle of 120°; find the ratio of the weights at the ends of the arms when the lever is in equilibrium with one arm horizontal.

13. ACB is a bent lever with its fulcrum at C, the angle ACB is a right angle, the arms AC and BC are 10ft. and 7ft. long, and AC is in a vertical position; a horizontal force of 21lbs. acting at A, is balanced by a vertical force P, acting at B; find the magnitude of P and the pressure on the fulcrum, and show by a diagram the line along which the latter acts.

SECTION III.

Equilibrium of a body acted on by any number of Forces.

Ex. 4. A uniform ladder whose weight is W, and may be supposed to act at its middle point, rests with its lower end upon a smooth horizontal plane, and its upper end on a slope inclined at an angle of 60° to the horizon; the ladder makes an angle of 30° with the horizon; find the pressures on the plane and slope respectively; find also, the force which must act horizontally at the foot of the ladder to prevent sliding.

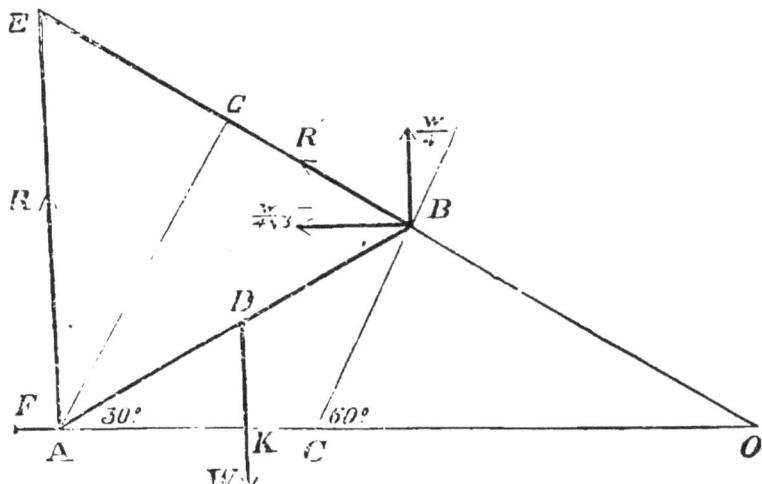

Let AC be the horizontal plane, CB the slope, and AB the ladder.

Let W = the weight of the ladder, acting at its middle point.

Let F = the force which prevents sliding.

,, R, R', = the pressures on the plane and slope respectively.

$2a$ = the length of the ladder.

From A draw AG perpendicular to direction of R'.

Then, taking moments about A, we have

$$R'.AG = W.AK \quad \ldots \ldots \ldots \ldots \ldots (1).$$

The angles ABG, ADK, each $= 60°$;

therefore $\quad AG = \dfrac{2a}{2}\sqrt{3}$;

and $\quad AK = \dfrac{a}{2}\sqrt{3}$.

Substituting the above values of AG and AK in (1), we have

$$R' a\sqrt{3} = W . \dfrac{a}{2}\sqrt{3};$$

$$\therefore R' = \dfrac{W}{2}.$$

Next, resolve R' vertically and horizontally, and we have $\dfrac{W}{4}$ and $\dfrac{W}{4}\sqrt{3}$ respectively.

Since there is equilibrium, we must have, Art. 61,

$$F = \dfrac{W}{4}\sqrt{3},$$

and $\quad R + \dfrac{W}{4} = W$;

$$\therefore R = \tfrac{3}{4} W.$$

We can obtain the same results by taking moments only.

Produce the directions of R and R' till they meet in E, and produce R' backwards till it meets the direction of F in O.

Then, ABE is an equilateral triangle, and therefore, $AE = AB = 2a$.

Taking moments about E, we have

$$F . AE = W . AK,$$

or $\quad F . 2a = W . \dfrac{a}{2}\sqrt{3}$;

$$\therefore F = \dfrac{W}{4}\sqrt{3}, \text{ as before.}$$

Again, since $AEO = 60°$, $AOE = 30°$, and since AE, the side opposite $30° = 2a$, the hypothenuse $OE = 4a$; and therefore, $AO = 2a\sqrt{3}$.

And $KO = AO - AK$

$$= 2a\sqrt{3} - \frac{a}{2}\sqrt{3}$$

$$= \frac{3a}{2}\sqrt{3}.$$

Taking moments about O, we have
$$R.\ AO = W.\ KO,$$
or $R.2a\sqrt{3} = W\frac{3a}{2}\sqrt{3}$;

$$\therefore R = \frac{3W}{4}, \text{ as before.}$$

EXERCISE III.

1. A uniform beam AB, 17 feet long, whose weight is 120 lbs. acting at its centre, rests with one end against a smooth wall, and the other end on a smooth floor, this end being tied by a string 8 feet long, to a peg at the bottom of the wall; find the tension of the string.

2. A beam AB rests with one end A against a smooth vertical wall, and the other end B on a smooth horizontal plane; it is prevented from sliding by a cord tied to one end of the beam and to a peg at the bottom of the wall; the length of the beam is 10 ft. 6 in., and the length of the string 9 feet. Suppose the weight of the beam to be 112lbs. and to act vertically through its middle point, find the forces acting on the beam.

3. Solve the first four questions under Exercise III, page 27, by the principle of moments.

4. A ladder, the weight of which is 90lbs., acting at a point one-third of its length from the foot, is made to rest against a smooth vertical wall, and inclined to it at an angle of 30°, by a force applied horizontally to the foot; find the force.

5. A uniform beam 6 feet in length, rests with one end against a smooth vertical wall, the other end resting on a smooth horizontal plane, and is prevented from sliding by a horizontal force applied to that end, equal

to the weight of the beam, and by a weight equal to two-thirds the weight of the beam, suspended from a certain point on the beam. Find the distance of this point from the lower end of the beam, the beam being inclined at an angle of 45° to the horizon.

6. A roof ACB consists of beams which form an isosceles triangle, of which the base AB is horizontal. Given the weight of each beam 100lbs., and 45° the angle at which it is inclined to the horizon, find the force necessary to counterbalance the horizontal thrust at A.

7. A ladder the weight of which may be regarded as a force acting at a point one-third the length from the foot, rests with one end against a peg in a smooth horizontal plane, and the other end on a wall. The point of contact with the wall divides the ladder into parts which are as 1 : 4; having given that the ladder weighs 120lbs., and makes an angle of 45° with the horizontal plane, find the pressure on the peg, and the reaction of the wall.

Ex. 5. AB is a string, whose length is 20 feet, and BC a pole, whose length is 13 feet, and weight 10lbs. acting at its centre; together they support a weight of 160lbs; find the tension of the string, if AC = 11 ft.

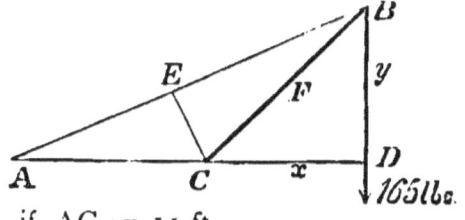

Let $CD = x$, and $BD = y$;
then $AB^2 = (AC + x)^2 + y^2$,
and $BC^2 = x^2 + y^2$.
Hence $AB^2 - BC^2 = AC^2 + 2 AC. x$,
or $20^2 - 13^2 = 11^2 + 22x$;
therefore $x = 5$.
And $13^2 = 5^2 + y^2$;
$\therefore y = 12$.

The 10lbs at F, the centre of BC, may be resolved

into 5lbs. at C and 5lbs at B; the latter may be added to the 160lbs.

Draw CE perpendicular to AB.
then $20 \times CE = 2$ area triangle ABC,
and $11 \times 12 = 2$ area of same triangle;
hence $20 \times CE = 11 \times 12$;
$\therefore CE = 6\frac{3}{5}$.

Let T be the tension of the string AB.
Taking moments about C, we have
$$T \times 6\frac{3}{5} = 165 \times 5;$$
$$\therefore T = 125 \text{lbs}.$$

EXERCISE IV.

This exercise requires a knowledge of easy deductions from the propositions of Euclid Book I.

1. ABC is an isosceles triangle, and D any point in the base BC. If equal forces act along the sides AB, CA, prove that the sum of the moments about D is independent of the position of D.

2. Three equal forces act in order along the sides of an equilateral triangle; show that the sum of the moments about any point within the triangle is invariable. If the point be outside the triangle, show that the algebraic sum of the moments is invariable.

3. Show by the Principle of Moments that if three forces act in consecutive directions round a triangle ABC, and be represented by its sides, they are not in equilibrium. Show also that if two forces act in consecutive directions along two sides of a triangle, and be represented by them, no force acting along the third side and represented by it, can produce equilibrium.

4. One end of a uniform beam is placed on the ground against a fixed obstacle, and to the other is attached a string, which runs in a horizontal direction to a fixed point vertically above the obstacle, and, passing freely over it sustains a weight W at its extremity, the beam being thus held at rest at an inclination of 45° to the horizon: prove that, if the string were attached to the

centre instead of the end of the beam, and passed over the same fixed point, a weight at the end of the string, equal to $W\sqrt{2}$, would keep the beam at rest in the same position.

5. At what point of a tree must a rope of given length a be fixed, so that a man pulling at the other end may exert the greatest force in upsetting it?

CHAPTER VII.

CENTRE OF GRAVITY.

SECTION I.

Definition; Centre of gravity of a uniform straight rod, and of bodies lying in the same straight line.

65. Introduction. The attraction of the earth, which causes a body to have weight, acts on every particle of the body; if a stone, for example, be broken into small fragments, the sum of the weights of the particles will be equal to that of the whole body. If one of these particles be attached by a fine thread to a fixed point, the thread will take the direction of the vertical through the point, and if several of them be suspended from points near together, the threads will be parallel. When, therefore, the particles are united so as to form the body, we may regard their weights *as a system of parallel forces*.

Regarding the earth as a sphere, it is true that the vertical lines would all converge to its centre, and therefore, strictly speaking, the direction of the forces which the earth exerts on the different particles composing a body are not parallel. But since the dimensions of any body we have to consider are very small compared with the radius of the earth, we may consider these directions to be appreciably parallel. For a similar reason we may suppose that the weight of any body is the same in whatever position it may be placed. The resultant of this system of parallel forces is the weight of the body.

It has been shown (Art 50), that the resultant of a system of parallel forces acting on a rigid body passes through a fixed point, the position of which is independent of the direction of the forces; the point at which this resultant acts is called the *centre of gravity* of the body. We have, therefore, the following definition:—

66. *Definition of Centre of Gravity.* **The centre of gravity of a body is the centre of parallel forces due to the weights of the respective parts of the body.**

67. If Centre of Gravity be fixed the body will rest in every position. If the resultant of the forces acting on a body be equal to the weight of the body, and act vertically upwards through the centre of gravity, it is evident from the definition that the body will be at rest. This statement is sometimes taken as the definition of the Centre of Gravity, and expressed thus:—

The Centre of Gravity of a system of heavy particles is a point such that, if it be supported and the parts rigidly connected with it, the system will rest in any position.

68. Centre of Gravity of a uniform straight rod. The centre of gravity of a uniform straight rod is at its middle point. For we may suppose the rod to be made up of an indefinitely large number of equal particles. Take two of these which are equidistant from the middle point of the rod; their centre of gravity is at the middle point. And since this is true for every such pair of particles, the centre of gravity of the whole rod is at the middle point of the rod. Hence, the weight of a uniform straight rod may always be supposed to be collected at its middle point.

EXAMPLES.

1. On a uniform straight lever weighing 5lbs., and 5ft. in length, weights of 1, 2, 3, 4lbs., respectively, are hung at distances 1, 2, 3, 4ft., respectively, from one end, find the centre of gravity of the system.

$$A \underset{G}{\rule{4cm}{0.4pt}} \overset{B \quad C \quad D \quad E}{\rule{0pt}{0pt}} F$$

Let AF be the lever, and let G be its centre of gravity. The whole weight of the lever may be supposed to act at G.

Let the given weights act at the points B, C, D, E, respectively.

Let x = the distance of the C.G. of the whole system from A.

The resultant of all the forces which act on the lever is the sum of the weights, together with the weight of the lever, and acts at the centre of gravity of the system.

Take moments about A; and since the moment of the resultant about any point is equal to the sum of the moments of the forces about the same point, we have

$$15x = 1 \times 1 + 2 \times 2 + 5 \times 2\tfrac{1}{2} + 3 \times 3 + 4 \times 4$$
$$= 42\tfrac{1}{2};$$
$$\therefore x = 2\tfrac{5}{6}\text{ft.}$$

The C.G. is, therefore, 2ft. 10in. from A.

2. A uniform beam AB, 20ft. long, is suspended from a nail by a string which is fastened to the beam at a distance of 2ft. from its centre; a weight of 20lbs. is attached to the other end to keep the beam horizontal. What is the weight of the beam?

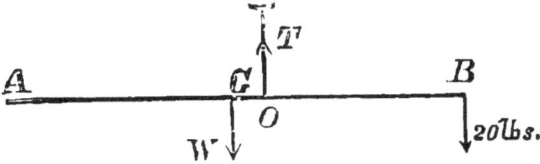

Let G be the C.G. of the beam; then W, the weight of the beam, may be supposed to act at G.

Let T be the tension of the string, and O the point in the beam to which it is attached.

The beam is kept at rest by three forces W, T, and 20lbs.; T is evidently equal to the resultant of the other two.

Take moments about O; and since the moments about any point in the direction of the resultant are equal and opposite (Art. 56), we have

$$W \times OG = 20 \times OB,$$
or
$$W \times 2 = 20 \times 8;$$
$$\therefore W = 80 \text{lbs}.$$

EXERCISE I.

1. Find the centre of gravity of 12lbs. and 20lbs. respectively, the line joining their centres of gravity being 2ft. 8in.

2. Three weights of 1lb., 2lbs. and 3lbs., are placed a foot apart, along a straight line; find their centre of gravity.

3. Two weights of 6lbs. and 12lbs. are suspended at the ends of a uniform horizontal rod, whose weight is 18lbs., and length 2ft.; find the centre of gravity.

4. Weights of 2lbs., 4lbs., 6lbs. and 8lbs., are placed so that their centres of gravity are in a straight line, and six inches apart; find the distance of their common centre of gravity from that of the larger weight.

5. Three weights of 4lbs., 6lbs. and 8lbs., respectively, are placed at intervals of 9in. along a weightless rod; find the distance of the centre of gravity from the middle of the rod.

6. A bar of uniform thickness and density, and 4 feet in length, has a weight of 10lbs. attached to one end; it balances about a point 9 inches from that end. What is the weight of the bar?

7. A bar of uniform thickness and density, and weighing 5lbs., has a weight of 10lbs. attached to one end, and a weight of 12lbs. is suspended from the other; it balances about a point 4 inches from the middle. What is the length of the bar?

8. A uniform stick 6 ft. long lies on a table, with one end projecting beyond the edge of the table to the extent of two feet; the greatest weight that can be suspended from the end of the projecting portion without destroying the equilibrium is 1lb.; find the weight of the stick.

9. Four weights of 3lbs., 2lbs., 4lbs., and 7lbs., respectively, are at equal intervals of 8 inches on a lever without weight, two feet in length; find where the fulcrum must be in order that they may balance.

10. A ladder 20 ft. long weighs 60lbs.; its centre of gravity is 8 ft. from its thicker end; it is carried by two men, one of whom supports the heavier end on his shoulder; where must the other stand that the weight may be equally divided?

11. A heavy tapering rod, having a weight of 20lbs. attached to its smaller end, balances about a fulcrum placed at a distance of 10ft. from the end; the weight of the rod is 200lbs.; find the point about which it will balance when the attached weight is removed.

12. A uniform bar of iron 10ft. long, projects 6ft. over the edge of a wharf, there being a weight placed upon the other end; and it is found that when this is diminished to 3 cwt. the bar is just on the point of falling over; find its weight.

13. A cylindrical vessel weighing 4lbs., and the internal depth of which is 6 in. will just hold 2lbs. of water. If the centre of gravity of the vessel when empty is 3.39in. from the top, determine the position of the centre of gravity of the vessel and its contents when full of water.

14. Prove that if the centre of gravity be found for one position of a body, it will be the centre of gravity when the body is turned into any other position.

15. What hypotheses not realized in nature, are made in the definition of the centre of gravity?

SECTION II.

Properties of the Centre of Gravity.

69. *Every body has a centre of gravity.*

For, if it be divided into an indefinite number of small particles, these particles are the points of application of a system of parallel forces equal to their weights; the centre of gravity of two of these may be found by the method already given; join this point with a third particle and find the centre of gravity of the sum of the first two particles and the third, and so on to the last particle. The point thus found will be the centre of gravity.

70. *No body can have more than one centre of gravity.*

For, if it be possible, let a body have two centres of gravity, and place it so that the line joining them shall be horizontal; then since the resultant of the weights of the several particles of the body passes through each of these centres, its direction must be the right line joining them, that is to say, a horizontal line, which is absurd. Therefore the body cannot have two centres of gravity.

71. *When a body is suspended from a point about which it can turn freely, it will rest with its centre of gravity in the vertical line passing through the point of suspension.*

For the body is acted on by two forces, viz., its own weight in a vertical direction through the centre of gravity, and the force arising from the fixed point. The body will not rest unless these two forces are equal and opposite. Therefore, the centre of gravity must be in the vertical line which passes through the point of suspension.

72. Experimental method of finding centre of gravity. The preceding article suggests an experimental method of finding the centre of gravity of a body. Let the body be suspended from two points successively, and let the vertical line through the point of suspension in each position be marked upon it. Each of these lines must pass through the centre of gravity, therefore, their intersection is the point required.

CENTRE OF GRAVITY.

73. A body placed on a horizontal plane will stand or fall according as the vertical through its centre of gravity falls within or without the base.

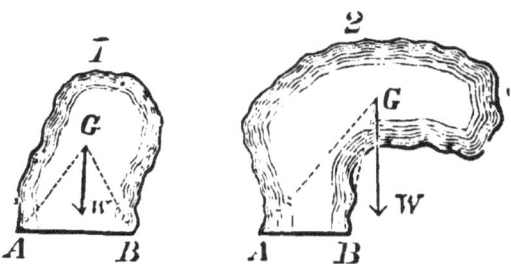

Let G be the centre of gravity of a body resting on a horizontal plane. The weight of the body, W, is a vertical force acting through G. If the body fall over, it must turn about a line touching the base. In (1) this rotation will raise G, and therefore, cannot be produced by gravity alone ; but in (2) this rotation will cause G to descend, and the tendency of gravity is to bring G lower if possible, or to induce the motion of falling. Hence in (1) the body will stand, and in (2) it will fall over. If G be vertical above A or B the body will stand, but the slightest disturbance will upset it. If the plane be inclined, the vertical through G must still fall within the base, or equilibrium will be impossible.

74. Def. of base. By the term base in the preceding article, is meant the area enclosed by a string drawn tightly round the points of support of the body ; thus, if the body stand on three legs, by joining these a triangle will be formed and this will be the space within which the vertical line from the centre of gravity must fall.

75. Remarks. A man must stand in a vertical position, in order that the centre of gravity may fall within the limits of his feet, which form the base on which he stands. This base may be enlarged by separating the feet, and the man's steadiness is correspondingly increased. If a person raise one foot from the ground, then his base is reduced to the sole of his foot and his steadiness is diminished. Men, and indeed all animals acquire the habit of instinctively shifting their position, so as to satisfy this condition of equilibrium; thus, if a man walking upon a narrow plank feels himself in danger of falling upon one side, he throws out the opposite arm; a man carrying a bundle upon his back leans forward; in walking down a hill, we lean backwards; in rising from a chair we must either lean forward to bring the centre of gravity over the feet, or else we must put the feet backwards under the chair to produce the same effect.

76. **Stable Equilibrium.** A body is said to rest in Stable Equilibrium, when, upon being disturbed in a very slight degree from its position of equilibrium, it will, upon the disturbing cause being withdrawn, return to its first position.

Unstable Equilibrium. If however, the body tend to move further from its original position, that position is called one of unstable equilibrium.

Neutral Equilibrium. If it remain in the new position, which the displacement has given it, the position is said to be neutral.

A weight suspended by a string from a fixed point is an example of stable equilibrium, for if it be slightly pulled out of its position, it will tend to return to its original position.

A pencil balanced on the end of the finger is in unstable equilibrium, for if it be in the least degree disturbed, it will fall away from the finger.

A sphere resting on a horizontal table, will remain in its new position if slightly disturbed, and is therefore in neutral equilibrium.

EXERCISE II.

1. Show that a body has one centre of gravity and only one?

2. Why does a person carrying a heavy weight in his hand lean towards the opposite side?

CENTRE OF GRAVITY.

3. Explain under what circumstances a body placed on a horizontal plane will remain at rest.

4. Explain how the centre of gravity of a body may be experimentally determined, by suspending it from a point.

5. When a body is placed on a horizontal plane, it will stand or fall, according as the vertical line, drawn from the centre of gravity, falls within or without the base. Explain fully what you mean by the term base in this proposition, and give familiar examples.

6. How is it that an inclined tower, such as that of Pisa, does not fall, although its top hangs about twelve feet over the base?

7. Supposing such a tower to be built in the form of an oblique cylinder, so that the slant side is to the height as $5 : 4$; what is the greatest height to which it may be built in theory, if the radius of the base be 30 feet?

8. Find the height of a cylinder which can just rest on an inclined plane, the angle of which is $60°$; the diameter of the cylinder being 6 inches.

9. Define the terms *stable*, *unstable*, and *neutral* equilibrium. Give familiar examples.

10. If a body be in stable equilibrium, how is the centre of gravity affected by a small displacement of the body?

11. A body cannot be in stable equilibrium upon a horizontal plane, if it rests on less than three points of support.

12. Why is a solid cylinder resting on a horizontal plane more difficult to upset, than a hollow cylinder of the same dimensions?

13. Why is it necessary that a table should have three legs at least?

14. Why cannot a pin practically be made to stand upon its point?

15. Why is it more dangerous to place luggage on the top of a coach, than in the body?

16. Two carriages have the same base, but the centre of gravity of the one is higher than the centre of gravity of the other; which of the two is more easily upset? Illustrate by a diagram.

17. Two books similar in every respect, each 10 inches long, lie one exactly upon the other on a table, over the edge of which they project 3 inches. How much farther may the upper book be pushed out before they fall over?

18. A body is in shape a sphere, but loaded in such a manner that its centre of gravity is not at its geometrical centre; when it is placed on a horizontal plane, what are its positions of stable, and unstable equilibrium?

19. Explain how a long rod is more easily balanced or its end than a short one.

20. An equilateral triangle is placed upon an inclined plane, its lowest angle being fixed; find how high the plane may be elevated before the triangle rolls.

21. A right-angled triangular board, hangs at rest from the right-angle, and the hypothenuse is inclined at 60° to the plumb line; compare the lengths of the sides.

SECTION III.

Centre of Gravity of plane areas, &c.

77. *To find the centre of gravity of a triangle.*

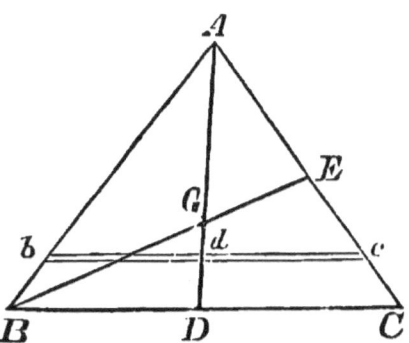

Let ABC be the triangle which is supposed to be of small and uniform thickness.

We may conceive this triangle to be made up of a number of rods all parallel to BC. Let *bc* be one of the series of rods; the centre of gravity of *bc* will be *d* its middle point (Art. 68); and *d* will be a point on the line AD drawn from A to the bisection of BC. And the centre of gravity of every other one of the series of rods must also lie on this line AD. Therefore the centre of gravity of the whole triangle must be somewhere in the line AD.

Now we might have supposed the triangle to have been made up in the same way of a series of rods all parallel to the side AC, and then we should have obtained the result that the centre of gravity of the whole triangle must be somewhere in the line BE, drawn from B to the bisection of the side AC.

Since, therefore, the centre of gravity of the triangle is in the line AD, and also in the line BE, it must be the point G where these two lines intersect.

78. Distance of C.G. from middle point of base. We may find the value of FG, or EG, as follows:

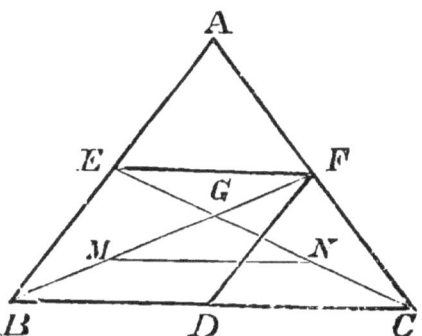

Bisect AB, AC, BC, BG, CG, in E, F, D, M, and N, respectively.

Then since AC is bisected in F,

the triangle BFC $= \tfrac{1}{2}$ the triangle ABC (Euc. I. 38).
Similarly ,, CEB $=$,,   ABC.
Therefore ,, BFC $=$ the triangle CEB,
and therefore EF is parallel to BC (Euc. I. 39).

In a similar manner it may be shown that FD is parallel to AB.

Therefore BDFE is a parallelogram.

Hence EF is equal and parallel to BD (Euc. I. 34).

Similarly MN "   "   BD;

therefore MN "   "   EF.

Hence the angle EFG = the angle GMN (Euc. I. 29);

and "   EGF = "   MGN (Euc. I. 15).

And the side EF has been shown to be equal to the MN.

Therefore FG = GM
$$= MB, \text{ because BG is bisected in M,}$$
$$= \tfrac{1}{3} BF.$$

Similarly it may be shown that
$$EG = \tfrac{1}{3} CE.$$

Hence, to find the C. G. of a triangle ABC: *from an angle draw a line to the bisection of the opposite side; the centre of gravity lies on that line, and at a point one-third its length from the side which it bisects.*

Hence also, the right lines drawn from the angles of a triangle to bisect the opposite sides all pass through the same point, for they all pass through the centre of gravity.

79. **Remarks on C. G. of plane areas.** When we speak of the centre of gravity of a triangle, it would be more correct to speak of the centre of gravity of a portion of matter bounded by two plane triangles, the surfaces of which are parallel and very near to each other, or of the centre of gravity of a very thin lamina on a triangular base. No confusion can arise from speaking of the centre of gravity of a plane figure, if the student bears in mind that his results are applicable to indefinitely thin plates or lamina; or if thickness be considered he must regard the centre of gravity as lying inside the body and at equal distances from the two bounding surfaces.

CENTRE OF GRAVITY.

80. *Heavy particles are placed at the angles of a triangle, their weights being proportional to the sides opposite to them: to find their centre of gravity.*

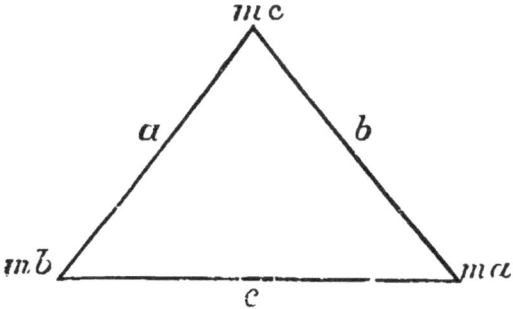

Let a, b, c, be the sides of the triangle, and ma, mb, mc, the weights at the opposite angles.

Let x, y, z, denote the distances of the centre of gravity of the weights from the sides a, b, c, respectively; and let p be the perpendicular upon the side a from the opposite angle.

Take moments about the side a, and we have
$$(ma + mb + mc)\,x = ma\,p\,;$$

therefore
$$x = \frac{ap}{a+b+c}$$
$$= \frac{2A}{a+b+c},$$

where A is the area of the triangle.

Similarly
$$y = \frac{2A}{a+b+c};$$

and
$$z = \frac{2A}{a+b+c}.$$

These results show that the centre of gravity of the three weights is equally distant from the three sides of the triangle. This point is the intersection of the lines bisecting any two angles of the triangle.

81. *To find the centre of gravity of the periphery of triangle formed by a piece of uniform wire.*

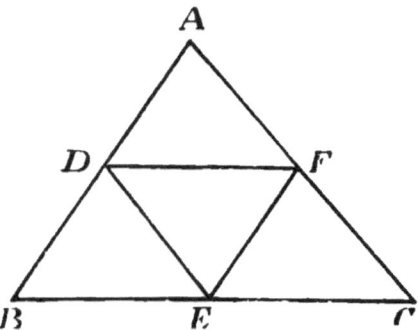

Bisect the sides of the triangle ABC in D, E and F. The weight of each side may be supposed to be collected at its middle point, it is also proportional to the side, and therefore proportional to the opposite side of the triangle DEF; hence the problem becomes the same as the one just solved.

82. *To find the centre of gravity of a parallelogram.*

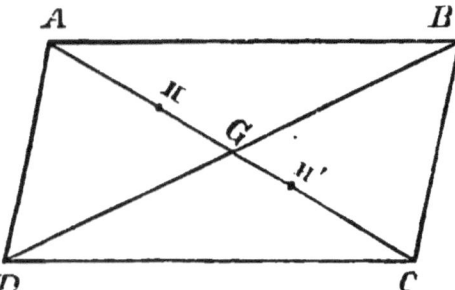

Let ABCD be any parallelogram; draw the diagonals AC, BD, intersecting in G. These diagonals mutually bisect each other. The centre of gravity H of the triangle DAB is in the line AG at a distance $= \frac{1}{3}$ GA from G (Art. 78); and the centre of gravity H' of the triangle BCD is at a distance $= \frac{1}{3}$ GC from G; also the triangles are both equal, and consequently their weights are so, therefore the resultant of the two weights is half-way between H and H'; or, in other words, the centre of gravity of the whole parallelogram is at G.

83. **Centre of Gravity of Symmetrical Figures**
If we can by inspection determine a point around which the material of a body is symmetrically situated, that point will be the centre of gravity of the body. It has been shown (Art. 68), that the centre of a uniform straight rod is its centre of gravity. This result might have been arrived at by the general consideration that around the centre of the rod, the material, being symmetrically situated, will be equally drawn towards the earth's centre; and therefore the centre of the rod must be the centre of the system of forces produced by the attraction of the different parts of the rod towards the earth. In the same way the centre of gravity of a sphere is evidently its geometrical centre, for any change in the position of the sphere can produce no change in the disposition of the material about its centre. Hence if the centre were fixed, the sphere would have no tendency to turn about that point, from one position into any other. The centre of gravity of the parallelogram, in the last article, might have been determined in this way; and in the same manner the centre of gravity of many geometrical figures may be found.

EXAMPLES.

3. At the corners of a square, taken in order, are placed weights 1, 3, 5, 7; find their centre of gravity.

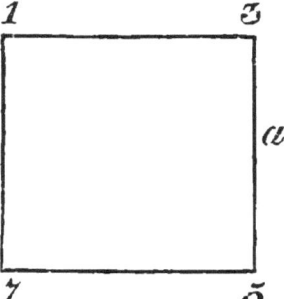

Let a, be the length of a side of the square.

Let $x =$ the distance of the centre of gravity from the side (1, 7); and let $y =$ the distance of the centre of gravity from the side (5, 7.)

The sum of the weights may be supposed to act at their centre of gravity.

Take moments about the side (1, 7), and we have
$$16x = 3a \times 5a;$$
therefore
$$x = \frac{a}{2}$$
Take moments about the side (5, 7), and
$$16y = 3a \times a;$$
therefore $y = \tfrac{1}{4}a.$

Hence, the distance of the centre of gravity from (1, 7) is $\tfrac{1}{2}a$, and from (5, 7), is $\tfrac{1}{4}a$.

4. Find the centre of gravity of a uniform circular disc out of which another circular disc has been cut, the diameter of the latter being the radius of the former.

Let R be the radius of the circle ABC,
" r " " AGO.

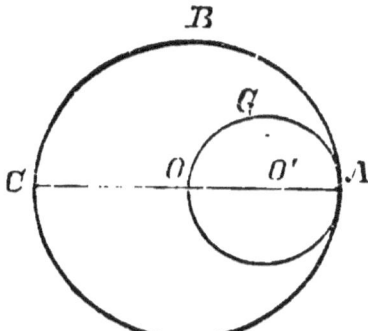

From the symmetry of the figure we see that the centre of gravity of the disc will lie on the diameter AC, which passes through the centres of both circles.

Let $x =$ the distance of the centre of gravity of the disc from O.

The weights of the circles are proportional to their areas and may be supposed to act at their centres of gravity which are their geometrical centres.

Take moments about O, and we have
$$\pi R^2 . 0 = \pi r^2 . r + (\pi R^2 - \pi r^2)x^*;$$
therefore $0 = r^3 + (R^2 - r^2)x$
$\qquad = r^3 + (4r^2 - r^2)x,$ since $R = 2r;$
$$\therefore x = -\frac{r}{3}$$
$$= -\frac{R}{6}$$

The negative sign shows that x lies to the left of O. Hence the centre of gravity of the disc lies between O and C and at a distance from O equal to $\tfrac{1}{6}$th the radius,

*If r be the radius of a circle, its area $= \pi r^2$, where $\pi = 3.1415 9 \ldots$

EXERCISE III.

1. At two of the angular points of an equilateral triangle are suspended weights of one pound, and at the third angular point is suspended a weight of two pounds; find the centre of gravity of the system.

2. Find the centre of gravity of three equally heavy particles placed at the three angles of a triangle, and show that their centre of gravity coincides with the centre of gravity of the triangle.

3. Find the centre of gravity of a right-angled isosceles triangle, and the squares described on the two equal sides.

4. A bent bar of uniform thickness and density, forms an equilateral triangle; show that the centre of gravity of the bar corresponds with the centre of gravity of the triangle.

5. Four heavy particles, whose weights are 4, 6, 5 and 3lbs. respectively, are placed in the corners of a square plate whose sides are 26 inches, and weight 8lbs.; required the distance of the centre of gravity from the centre of the plate.

6. From a circular plate, whose radius is $2a$, a circular piece, whose radius is a is cut away; the distance between the two centres is $\frac{1}{2}a$; required the distance of the centre of gravity of the remainder from the centre of the plate.

7. From a circular disc of radius R a circular disc of radius r, is removed, the distance between their centres being d; determine the centre of gravity of the remainder.

8. A round table whose weight is W, stands on three legs placed on the circumference at equal distances; what is the greatest weight which may be placed upon any part of the table without upsetting it?

9. A circular table, weighing 168lbs. is supported on four legs in the circumference which form a square; find the least weight which being at the circumference will overturn the table.

10. A triangular slab of uniform thickness is supported at its three angular points; show that the pressures on the supports are equal to one another.

11. An isosceles triangle is placed with its base, which is 2 ft. in length, upon a plane whose inclination is $30°$, and is prevented from sliding by a small obstacle placed at the lowest point of its base; what is the greatest height which the triangle can have without toppling over?

12. Three weights are placed in the angles of a right-angled triangle, and are proportional to the squares on the opposite sides; required the distance of the centre of gravity from the right angle.

13. Find the centre of gravity of three uniform beams forming a right-angled isosceles triangle, the greatest side of the triangle being 20 ft.

14. An isosceles triangle rests on a square, and the height of the triangle is equal to a side of the square; find the centre of gravity of the figure thus formed.

15. If a piece of uniform wire bent into the form of a triangle ABC, be suspended freely from the angular point A, prove that it will only rest with the side BC horizontal when the angle ABC is equal to the angle ACB.

16. A number of cent pieces are cemented together so that each just laps over the one below it by the *ninth* part of its diameter; find how many may be thus piled without falling, when the lowest stands on a horizontal plane.

17. A short circular cylinder of wood has a hemispherical end. When placed with its curved end on a smooth table it rests in any position in which it is placed. Determine the position of its centre of gravity.

18. Explain how a plane area is said to have a centre of gravity. Find the centre of gravity of two material plane isosceles triangles on the same base, their vertices being on opposite sides of the base at distances h_1, h_2, from it.

19. The centre of gravity of a body being given, and also that of a portion of it, show how to find the centre of gravity of the remainder.

20. State and prove the proposition which suggests a practical method of finding the centre of gravity of any plane area.

21. A quadrilateral is such that when it is suspended from any angular point the diagonal through the point of suspension is vertical; show that it must be a parallelogram.

CHAPTER VIII.

SIMPLE MACHINES OR MECHANICAL POWERS.

84. **Def. of Machine.** Any contrivance which enables us to change the point of application, direction, or intensity of a force may be called a *machine*.

85. **Object of machines.** The object of all machines, considered in a statical point of view, is to enable a certain force, as P, which is called the *Power*, to be in equilibrium with a second force, as W, which is termed the *Weight*. These distinctive names are given to the two forces because we are most familiar with the use of machines, when employed for the purpose of raising, or moving heavy bodies, by the application of a small force or power.

86. **Mechanical Advantage.** The *mechanical advantage* of a machine may be defined as the ratio of the weight to the power when there is equilibrium; thus if a weight of 5 pounds sustains a weight of 80 pounds, the mechanical advantage is 80 ÷ 5 or 16. This value, $\frac{W}{P}$, is also the measure of the *efficiency* or *working power* of a machine.

When W is greater than P the machine is said to work at a *mechanical advantage*, when W is less than P at a *mechanical disadvantage*.

87. Mechanical Powers. The following six machines are, for convenience, regarded as simple machines or Mechanical Powers.
1. The Lever.
2. The Wheel and Axle.
3. The Pulley.
4. The Inclined Plane.
5. The Wedge.
6. The Screw.

Every machine, however complicated, may be shown to be composed of combinations of these simple machines.

THE LEVER.

I.

88. Kinds of Levers. We have already considered the principle of the lever as a general mechanical principle, and we have shown that two forces will balance about a fulcrum when the moments about it are equal; but the lever may also be regarded as one of the Mechanical Powers, and so considering it we distinguish three kinds of levers according to the position of the fulcrum with respect to the power and the weight.

A lever of the *first* kind has the fulcrum between the power and the weight. In this case any amount of mechanical advantage may be gained by making the arm upon which the power rests sufficiently long. A crowbar used to lift great weights, a poker, a pair of scissors, are examples. In the poker, the coals are the weight, the bar of the grate the fulcrum, and the force exerted by the hand, the power.

A lever of the *second* kind has the weight between the fulcrum and the power. The oar of a boat is an example, in which the water forms the fulcrum, the resistance of the boat, applied at the rowlock, is the weight, and the power is applied by the hand of the rower.

A lever of the *third* kind has the point of application of the power between the fulcrum and the weight. The most interesting example is the human arm when applied to lift a weight by turning about the elbow; here the elbow-joint is the fulcrum, and the power is applied by means of the muscles at a point between the joint and the weight. A man lifting a long ladder with one end resting on the ground is also a familiar example.

89. Remarks. These distinctions are, however, of but little practical importance. The turning effect of a force is measured by its moment; for equilibrium, therefore, the moment of the power must be equal to the moment of the weight. When this one idea is grasped and the position of the fulcrum ascertained the subject-matter is exhausted.

EXERCISE I.

1. It is said that the "six simple machines are the lever, the wheel and axle, the pulley, the inclined plane, the wedge, and the screw." Point out that some of these machines are merely modified from others.

2. When does a force act by a machine at a mechanical advantage? What is meant by mechanical advantage being lost or gained by the intervention of a lever? Explain under what circumstances either the one result or the other takes place in the case of each kind of lever.

3. There are three classes of levers; what distinguishes one class from another? Give familiar examples of each class describing it so as to show clearly that it is an example. To what class does a wheelbarrow belong?

4. How does the contrivance of placing the row-locks outside the boat affect the efforts of the rower?

5. "We have spoken of only two forces, the power and the weight, as acting upon the lever, yet in reality there must be three forces." What is the third force?

6. State the conditions for equilibrium of pressures acting on a lever.

7. If a walking-stick be supported near one end by passing that end over the thumb and fourth finger, it is noticeable that the pressure perpendicular to the stick which must be exerted by the fourth finger is greater

when the stick is held horizontally than when it is any slanting position. Explain the fact clearly on mechanical principles, with the help of a diagram.

8. If the forces acting perpendicularly at the extremities of the arms of a lever, balance each other, they are *inversely* as the arms.

9. Show *why* a lever is in equilibrium when the power and the weight are to each other inversely as the perpendicular distances of their lines of action from the fulcrum.

10. A man wheels a wheelbarrow along a level road. Point out the conditions which determine how much of the total weight of the load and barrow is supported by the wheel and how much is supported by the man.

SECTION II.

Balances.

90. Introduction. The lever is a convenient instrument for ascertaining the weight of a body by comparing it with another whose weight is known. The weights of the body to be weighed, and of the counterpoise with which it is compared, are two parallel forces acting in the same direction, and may therefore be made to balance each other by means of a lever of the first kind. Let W be the body to be weighed, and P the counterpoise, then if they do not balance one another, equilibrium may be produced (1) by changing the counterpoise P, (2) changing its position, (3) changing the place of the fulcrum, or (4) changing the inclination of the arms of the lever to the direction of the forces.

91. The Common Balance. The first of these methods is used in the balance with equal arms, usually called the *common balance.* In this balance the position of the fulcrum and of the points of suspension of the weight and its counterpoise are fixed, and the balance is so constructed, that it will be horizontal when the weight and its counterpoise are equal. For this purpose it is necessary that the arms of the lever should be equal, and also that the centre of gravity of the instrument itself should be in the perpendicular from the fulcrum to the line joining the points of suspension of the scales. It is moreover necessary, that the centre of gravity of the balance and its load, should always lie below the fulcrum, in order that, if the instrument were displaced, it may return to the horizontal position, for if the centre of gravity lay above the fulcrum, it would have a tendency to upset, and if it coincided with the fulcrum, it would rest indifferently in any position.

92. Qualities of a good Balance. The most important qualities of a good balance are:—

1. *Sensibility.* The beam should be sensibly deflected from a horizontal position by the smallest difference between the weights in the scale-pans. The definite measure of the sensibility is the angle through which the beam is deflected by a stated percentage of the difference between the loads in the pans.

2. *Stability.* This means rapidity of oscillation, and consequently speed in the performance of a weighing. It depends mainly upon the depth of the centre of gravity of the whole below the knife-edge, and the length of the beam.

3. *Constancy.* Successive weighings of the same body must give the same result—all necessary corrections depending on temperature, &c., being allowed for.

93. Remarks. The *sensibility* of a balance is, in general, of more importance than the *stability*, but much depends on the service for which the balance is intended. For weighing heavy goods stability is of more importance. When great accuracy is required, as in chemical balances, sensibility is the quality desired.

94. *To determine the true weight of a body by a false balance.*

Let W be the true weight of the body, and let x and y be the unknown arms of the balance. When W is placed in one scale-pan, let it be balanced by a weight

P, and when placed in the other by a weight Q. Then since W and P acting at the arms x and y, were in equilibrium, we have
$$Wx = Py \dots \dots \dots \dots (1).$$

And since W and Q, acting at the arms x and y, were in equilibrium, we have
$$Wy = Qx \dots \dots \dots \dots (2).$$
Multiplying (1) and (2), $W^2xy = PQxy$.
Or $\quad\quad\quad\quad W^2 = PQ$;
$\therefore W = \sqrt{PQ}$.

That is, the true weight is the square root of the product of the false weights; or the true weight is a *mean proportional* between the two false weights.

EXAMPLES.

1. If a balance be false, having its arms in the ratio of 15 to 16; find how much per lb. a customer really pays for tea which is sold to him from the longer arm at 75 cents per lb.

Since the arms are in the ratio of 15 to 16, 16oz. at the end of the shorter arm will be balanced by 15oz. at the end of the longer arm. The customer will, therefore get 15oz. instead of 1lb. and for this he pays at the rate of 5 cents per oz. Therefore for 16 oz. he will pay 80 cents.

2. A merchant has correct weights but a false balance, that is one having one arm longer than the other; supposing that he serves out to each of two customers articles weighing, as indicated by his balance, W lbs., using for the commodities first one scale and then the other; find whether he gains or loses by the peculiarity of his balance.

Let a, b, be the lengths of the arms of the balance, and let P, Q, be the true weights of the commodities, placed at the end of the former and latter arm respectively,

Then $\quad\quad\quad Pa = Wb$,
and $\quad\quad\quad Qb = Wa$.

Hence $P = W \dfrac{b}{a}$;

and $Q = W \dfrac{a}{b}$.

Also $2W = 2W$;

$\therefore P + Q - 2W = W \left(\dfrac{b}{a} + \dfrac{a}{b} - 2\right)$

$= W \dfrac{a^2 + b^2 - 2ab}{ab}$

$= W \dfrac{(a-b)^2}{ab}$; a positive

quantity. The merchant therefore, loses by his sale to the two customers a weight of substance equal to $\dfrac{W(a-b)^2}{ab}$.

EXERCISE I.

1. How can you determine the true weight of an article by having it weighed in each scale successively of a common balance, that is rendered false,

(1) by having its arms unequal,

(2) by having one of its scales loaded?

2. If a straight lever balance, under the action of two weights, in one position, it will balance in every position.

What exceptions are there to this?

Why is it not the case in a common pair of scales?

3. If when a balance is suspended the beam be not horizontal, prove that if the want of horizontality arise from an inequality in the weight of the scale-pans, the balance may be corrected by putting a weight into the lighter of the two, but that if it arise from a difference of length of the arms the balance cannot be so corrected.

4. What is meant by the sensibility of a balance?

Other conditions being the same, why is the sensibility less, the greater the weight of the beam?

What are the requisites of a good balance?

5. A tradesman's balance has arms whose lengths are 11 in. and 12 in., respectively, and it rests horizontally, when the scales are empty. If he sells to each of two customers a pound of tea at 55 cents per ℔, putting his weights into two different scales for each transaction, find whether he gains or loses, owing to the incorrectness of his balance, and how much.

6. A common balance has its arms of unequal length; a body when placed in one scale-pan appears to weigh 4lbs., and when placed in the other 16lbs. What is the true weight of the body?

7. The whole length of the beam of a false balance is 3ft. 9in.; a certain body placed in one scale appears to weigh 9lbs, and placed in the other appears to weigh 4lbs. Find the true weight of the body, and the lengths of the arms of the balance, supposed to be without weight.

8. A grocer buys tea wholesale, at the rate of 80 cents per ℔., and in weighing it out to his customers uses a balance, the arms of which are in the ratio of 16 to 15; at what price must he profess to sell it per ℔., in order that he may make a profit of 20 per cent?

9. A tobacconist buys tobacco wholesale, at the rate of $66\frac{2}{3}$ cents per ℔., and in weighing it out to his customers uses a balance, the arms of which are in the ratio of 16 to 15; if he profess to sell it at 75 cents per ℔., what profit per cent. does he really make?

10. One of the arms of a false balance is longer than the other by $\frac{1}{4}$ part of the shorter: when used the weight is put into one scale as often as into the other. What will be the gain or loss per cent. to the seller?

11. The arms of a false balance are to one another as 7 to 8, and the weights are put into one scale as often as into the other; what will be the gain or loss per cent. to the seller?

THE ROMAN STEELYARD.

95. The Roman Steelyard. The Common or Roman Steelyard is another form of the balance. It is represented in the annexed figure.

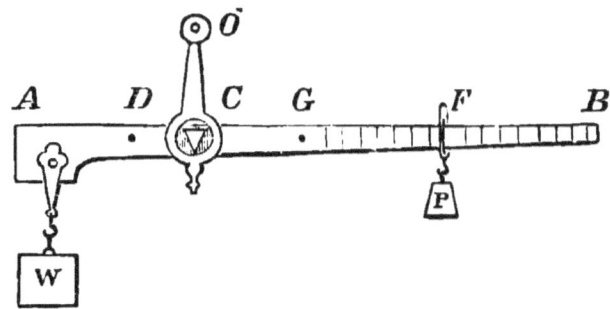

AB is an iron bar turning about the edge C of a triangular piece of metal, which works in an aperture of the handle by which the balance is suspended from a point O.

At A, a hook or scale is attached for the purpose of receiving any article whose weight is to be ascertained. P is a movable weight which can be suspended from any point between C and B. Suppose W to be a body suspended from A, then P must be moved toward C, or towards B, until the beam AB rests in a horizontal position. Let F be this position of P. The bar is so graduated from C to B that the reading of F indicates the weight of W.

96. *To graduate the Common Steelyard.*

Let the weights W and P be removed.

Suppose G to be the centre of gravity of the beam.
" W' " weight of the beam.
The arm CB being by construction heavier than the rest of the beam, will fall when the machine is in this unloaded state.

Let the weight P be placed on the other side of the handle CO at a point D, found by trial, such that the beam is kept in a horizontal position. We shall then have

$$P.CD = W'.CG \dots \dots \dots (1).$$

Now let the beam be loaded with P and W as before; then for equilibrium we must have

$$P \cdot CF + W' \cdot CG = W \cdot AC.$$

Substituting from (1), we have

$$P(CF + CD) = W \cdot AC;$$

$$\therefore W = P \cdot \frac{DF}{AC} \dots (2).$$

Suppose now we put W, 1lb., 2lbs., 3lbs.,.......... successively, we can calculate in each case the corresponding position of F from equation (2). At the point in the arm so determined graduations must be made and marked 1, 2, 3, &c.

The space between two graduations may be subdivided into four or more equal parts. and these subdivisions will indicate with sufficient accuracy, the alteration which must be made in W to indicate P placed at any of them.

EXERCISE II.

1. Describe the common (Roman) Steelyard, and investigate the method of graduating it.

2. Where must be the centre of gravity of the common steelyard so that any movable weight may be used with it?

3. If the movable weight for which the steelyard is constructed be 1lb., and a tradesman substituted 2lbs., using the same graduations, show that he defrauds his customers, if the centre of gravity of the steelyard be in the longer arm, and himself if it be in the shorter arm.

4. The movable weight is 1lb., and the weight of the beam is 1lb.; the distance of the point of suspension from the body weighed is $2\frac{1}{2}$ inches, and the distance of the centre of gravity of the beam from the body weighed is 3 inches; find where the movable weight must be placed when a body of 3lbs. is weighed.

5. If the fulcrum divide the beam, supposed uniform, in the ratio of 3 to 1, and the weight of the beam be equal to the movable weight, show that the greatest weight which can be weighed is four times the movable weight.

6. Find what effect is produced on the graduations by increasing the movable weight.

7. Find what effect is produced on the graduations by increasing the density of the material of the beam.

THE WHEEL AND AXLE.

II

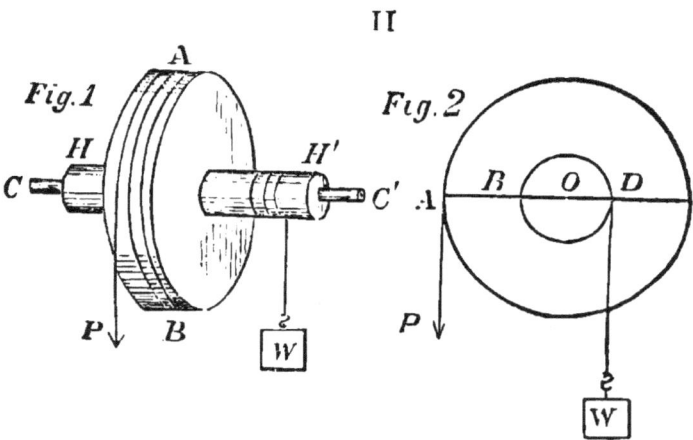

97. Description of Wheel and Axle. This machine consists of a wheel AB, and a cylinder or axle HH'; both turning round the same axis. The points C,C', are placed on bearings so as to allow of motion round the common axis. The power is applied at the circumference of the wheel, generally by a cord wrapped round it, as in the figure, and the weight is attached to a cord coiled round the axle in the contrary direction, so that when P descends, W ascends, and *vice versa*.

The second figure represents a vertical section of the machine, by a plane perpendicular to the axis. The power and the weight are, for the sake of simplicity, supposed to act in the same plane. O is the common centre. Draw through O the horizontal line ABOD. A and D will be the points where the cords, connected with the power and weight, leave the wheel and axle respectively.

98. Wheel and Axle a modification of the Lever. Suppose the figure to represent the machine at one moment of its action. We might, for that moment, suppose the whole machine reduced to a simple lever ABOD, of which O is the fulcrum and A and D the points of application of the power and weight respectively. We should not by this supposition affect in any way the forces at work. Hence we see the wheel and axle is an infinite number of levers, the common fulcrum being O, the arm at which the weight acts being always equal to the radius (OD) of the axle, and that at which the power acts equal to the radius (OA) of the wheel. Hence the wheel and axle is a practical arrangement for continuing the action of a lever as long as may be required, the weight rising all the time. On this account it is called the *perpetual lever*.

99. *To find the ratio of the Power to the Weight in the Wheel and Axle.*

Take moments about O (fig. 2, Art. 97), and we have
$$P \times AO = W \times OD,$$
or
$$\frac{P}{W} = \frac{OD}{AO} = \frac{r}{R};$$
where R is the radius of the wheel, and r the radius of the axle. There will, therefore, be equilibrium when

Power : Weight :: radius of Axle : radius of Wheel.

If the thickness of the ropes be considerable we must add half their thickness to R and r respectively.

EXERCISE I.

1. In the wheel and axle what is the difference between *axle* and *axis*? Is the axis fixed or rotatory?
2. Is there any advantage in having the rope which passes round the wheel thicker than that which passes round the axle?

3. Find the ratio of the power to the weight in the wheel and axle. Explain why the machine has been called the *perpetual lever*.

4. If the rope used be of such a thickness that it becomes necessary to take it into account what would be the ratio between P and W?

5. Why is the labour of drawing a bucket of water out of a common well generally greater during the last part of the process than during the first?

6. What is the greatest weight which can be supported by a power of 40lbs. by means of a wheel and axle, when the diameter of the wheel is 10 times that of the axle?

7. If a power of 35lbs. supports a weight of 84lbs. on a wheel and axle, and the radius of the axle be 5in., what must be the radius of the wheel?

8. Two men capable of exerting forces of 260 and 300lbs. respectively, work the handle of a winch and axis. The radius of the axle is 5in.; what must be the length of the arm of the winch that the men may be just able to raise a weight of 4480lbs?

THE PULLEY.

III.

100. **Def. of Pulley.** The Pulley consists of a small wheel which moves freely about an axis, and allows a string to pass over a groove on its surface. The ends of the axis are fixed to a frame called the *Block*.

101. **Def.** The Pulley is said to be *fixed* or movable according as the block is fixed or *movable*.

The wheel is supposed to revolve without friction, and the string to be perfectly flexible.

102. **Fixed Pulley.** No mechanical advantage is gained by a fixed pulley, for, as the tension in every part of the string is the same (Art. 15), if a weight W be suspended at one extremity, an equal weight must be applied at the other to maintain equilibrium; therefore, in this case,

$$P = W.$$

Hence, the effect of a fixed pulley is simply to change the direction of the force.

103. **Remarks.** The occasions upon which a single pulley is used are very numerous. Suppose a bag of wheat has to be raised from one of the lower to one of the upper stories of a building. It may, of course, be raised by a man carrying it; but in that case he has to carry his own weight in addition to the wheat. But if a rope be passed over a pulley at the top of the building he can raise the wheat without raising his own weight.

THE PULLEY.

104. The Single Movable Pulley.—*To find the ratio of the Power to the Weight in the single movable Pulley.*

Let the weight W be attached to the movable pulley B; and let B be sustained by the cord ABC, one extremity of which is fastened at A, and the other passing over the fixed pulley C sustains the power P.

The tension of the string is the same throughout. Hence we may regard the pulley as acted on by two parallel forces, each equal to P upwards, and by the force W downwards; therefore,

$$W = 2P.$$

The pressure on the fixed point A, is equal to P, that is to $\frac{1}{2}$W.

If the weight of the pulley is too considerable to be omitted, let it be equal to w; then we have

$$2P = W + w,$$
or $$P - w = \tfrac{1}{2}(W - w).$$

105. First System of Pulleys. *To find the ratio of the Power to the Weight, in a system of Pulleys, in which each Pulley hangs by a separate string.*

The annexed figure represents a system of pulleys in which the string that passes round one pulley has one of its ends fastened to a fixed point, and the other attached to the block of the next pulley; the weight is applied to the block of the lowest pulley, and the power to the extremity of the first string.

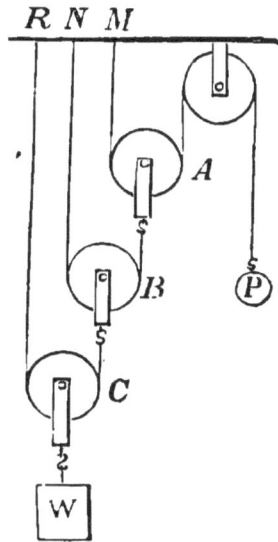

Let A, B, C, be three movable pulleys, and let P and W be in equilibrium.

By the principle of the single movable pulley, the tension of the string $BC = \tfrac{1}{2}W$,

" " $AB = \tfrac{1}{2}$ tension of $BC = \tfrac{1}{4}W$,
" " $AP = $ " $AB = \tfrac{1}{8}W$.
But " $AP = P$;
∴ $P = \tfrac{1}{8}W$.

Thus, when there are three movable pulleys,
$$P = \frac{W}{2^3}.$$

In the same manner it may be shown, that when there are n movable pulleys
$$P = \frac{W}{2^n},$$
or $2^n P = W.$

106. Pulleys supposed heavy. If the weight of the pulleys be taken into account, let the weight of the three pulleys, beginning with C, be w'_1, w'_2, w'_3, respectively;

the tension in
$$BC = \frac{1}{2}(W + w'_1),$$

" $$AB = \frac{1}{2^2}(W + w'_1) + \frac{1}{2}w'_2,$$

" $$AP = \frac{1}{2^3}(W + w'_1) + \frac{1}{2^2}w'_2 + \frac{1}{2}w'_3.$$

But " $AP = P$;

$$\therefore P = \frac{W}{2^3} + \frac{w'_1}{2^3} + \frac{w'_2}{2^2} + \frac{w_3}{2};$$

and similarly for any number of pulleys.

If the weight of each pulley be the same and equal to w, we have

$$2^3 P = W + w(2^3 - 1),$$

or $$P - w = \frac{1}{2^3}(W - w);$$

and similarly, if n be the number of pulleys,

$$P - w = \frac{1}{2^n}(W - w).$$

EXAMPLES.

1. In a system of pulleys, where each pulley hangs by a separate string, the number of pulleys is 5. What weight will a power of 5 lbs. support?

Here
$$W = 2^5 \times 5$$
$$= 32 \times 5$$
$$= 160 \text{ lbs.}$$

2. In the above system, 1 lb. is supported by 1 oz.; find the number of pulleys.

16 oz. $= 2^n \times 1$ oz., where n is the number of pulleys,

or $\quad 2^n = 16$;

$\therefore n = 4.$

THE PULLEY.

3. Four pulleys whose weights, beginning with the highest, are 3, 4, 2 and 6lbs., respectively, are arranged according to the first system; what power will sustain a weight of 462 pounds?

Here $$P = \frac{462}{2^4} + \frac{6}{2^4} + \frac{2}{2^3} + \frac{4}{2^2} + \frac{3}{2}$$
$$= \frac{462}{2^4} + \frac{50}{2^4}$$
$$= \frac{512}{16}$$
$$= 32 \text{lbs.}$$

4. In the case of the single movable pulley will the mechanical advantage be increased or diminished by taking into account the weight of the pulley?

When the weight of the pulley is neglected
$$\frac{W}{P} = 2.$$

But if the weight of the Pulley be taken into account,
$$2P = W + w,$$
or $$\frac{W}{P} = 2 - \frac{w}{P},$$ which is less than 2, and, therefore, the mechanical advantage is diminished.

EXERCISE I.

1. In the single movable pulley, is any mechanical advantage gained if the weight of the pulley be not less than that of the power?

2. In a system of four pulleys, where each pulley hangs by a separate string, the weight supported is 28lbs.; find the power.

3. Supposing a power of 3lbs. to sustain a weight of 48lbs., find the number of movable pulleys.

4. If a weight of 1lb. is supported by 1oz., what is the number of movable pulleys? Draw a figure to represent this case.

5. The mechanical advantage is equal to 16; how many pulleys are there?

6. If three movable pulleys, the weights of which are 2, 4, and 8 oz. respectively, beginning at the lowest, be arranged as in the first system, what is the least force that will raise a weight of 104lbs.?

7. With three movable pulleys arranged according to the first system, each of which weighs 8oz., what weight can be supported by a force of 1lb.?

8. In a system of pulleys, in which each pulley hangs by a separate string, there are three pulleys of equal weights; the weight attached to the lowest is 32lbs., and the power is 11lbs.; find the weight of each pulley.

9. In a system of five pulleys, in which each pulley hangs by a separate string, the tension of the string attached to the lowest block but one is 80lbs.; find the weight supported and the power exerted.

10. If a man weighing 160lbs. supports a weight equal to his own, and there are three pulleys, find his pressure on the floor on which he stands.

11. In the system of pulleys in which each pulley hangs by a separate string, a platform is suspended from the lowest block; what force must a man who weighs 140lbs., standing on the platform and pulling downwards on the string which passes round the highest pulley and over the pulley fixed to the beam, exert to sustain himself? when there are three movable pulleys

107. **Second System of Pulleys**. *To find the ratio of the Power to the Weight, in a system of Pulleys, in which the same string passes round all the Pulleys.*

In this system there are two blocks, each containing a number of pulleys. The lowest block is movable, the upper one fixed. The pulleys are of such relative sizes that the different portions of the string between them are all parallel.

This system will be understood from the annexed figure, and is known as the *Second System of Pulleys*.

Suppose there are four pulleys, two in each block. Let W denote the weight which is suspended from the lower block, and let P denote the power which acts vertically downwards at one end of the string. The tension of the string is the same throughout and is equal to P. We may, therefore, regard the lower block as acted on by four parallel forces, each equal to P upwards, and the force W downwards. Therefore

$$W = 4P.$$

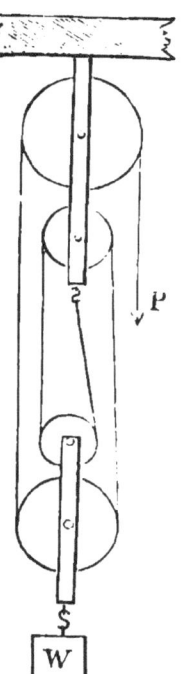

In a similar manner it may be shown that if there are n strings at the lower block,

$$W = nP.$$

If the weight of the lower pulleys and block be w, the total weight supported will be $W + w$, therefore

$$W + w = nP.$$

EXERCISE II.

1. In the system of pulleys where the same string passes round each pulley, if the number of strings at the lower block be six, what limit must be put to the weight of the lower block, so that any mechanical advantage may be gained by this system of pulleys?

2. In the system of pulleys where the same string passes round each pulley, there are 7 pulleys at the lower block. What weight will a power of 2lbs. support?

3. If there are twelve strings at the lower block in a system of pulley in which the same string passes round all the pulleys, find the weight which a power of 10lbs. will support, the weight of the pulleys being neglected.

4. If there are 6 strings at the lower block, in the Second System of pulleys, find the greatest weight which a man weighing 140lbs. can possibly raise.

5. In the Second System of pulleys the mechanical advantage is expressed by the number 10; how many movable pulleys are there?

6. Find the number of strings in the lower block in order that a power of 4oz. may support a weight of 4lbs.

7. A man weighing 168lbs. supports a weight equal to half his own; if there are 7 strings at the lower block, find his pressure on the floor on which he stands.

8. Find what weight can be supported by a power of 10lbs., if there are 3 pulleys at the lower block, and the weight of the lower block is three times the power.

9. In a system of pulleys where the same string passes round each pulley, if there are four pulleys at the lower block, two of which weigh 1lb. each, and the other two half a pound each, what weight will a power of 5lbs. support?

10. In a system of pulleys in which the same string passes round any number of pulleys, if the weight of the pulleys be regarded, under what circumstances will the mechanical advantage be reduced to nothing?

11. If six strings pass round the lower block and a man weighing 140lbs. supports himself by standing on a platform suspended from the lower block, and holding the rope that passes round the pulleys; find the tension of the rope.

12. A man sitting upon a board suspended from a single movable pulley pulls downwards at one end of a rope which passes under the movable pulley and over a pulley fixed to a beam overhead, the other end of the rope being fixed to the same beam. What is the smallest proportion of his whole weight with which the man must pull in order to raise himself?

13. With what force would he require to pull upwards, if the rope, before coming to his hand, passed under a pulley fixed to the ground, as well as round the other two pulleys?

THE PULLEY.

108. Third System of Pulleys. *To find the ratio of the Power to the Weight, in a system of Pulleys in which all the strings are attached to the weight.*

In this system the cord which passes over any one pulley has one extremity fastened to the block to which the weight is attached, while the other is attached to the block of the next pulley. The highest pulley is fixed.

This system is represented in the annexed figure and is known as the *Third System of Pulleys*.

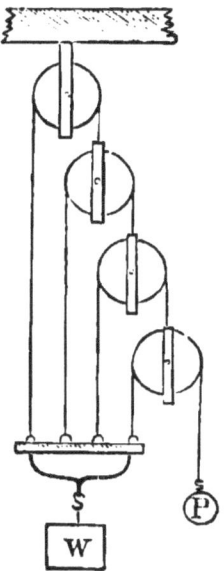

Suppose there are four pulleys. Let W denote the weight to which all the strings are fastened, and let P be the power which acts vertically downwards at the end of the string which passes over the lowest pulley.

The tension of the string which passes over the lowest pulley is P; the tension of the string which passes over the next pulley is $2P$; the tension of the string which passes over the next pulley is twice this or $2^2 P$; the tension of the string which passes over the highest pulley is twice this, that is, $2^3 P$. But the sum of all these tensions must equal W, since the strings support W;

therefore
$$W = P + 2P + 2^2 P + 2^3 P$$
$$= P(1 + 2 + 2^2 + 2^3)$$
$$= P(2^4 - 1).$$

Similarly, if n be the number of pulleys,
$$W = P(2^n - 1).$$

109. General Method of finding the relation of P to W in Pulleys.

Begin with the end of the system at which the Power acts. The tension of the string which supports P will be equal to P throughout; against each of the parallel portions of the string write P. Proceed to the next string, and find what its tension is by observing how many strings, each having the tension P, produce it; write the expression for its tension against each parallel portion; and so on with the next string. When the tension of each string of the system has been written down, it is easy to see how many of them support W, and by adding their tensions together we have the relation between P and W required.

EXAMPLES.

5. Find the power necessary to sustain a weight of 100lbs. with three movable pulleys arranged according to the Third System, the weights of the pulleys being 8oz., 6oz., and 4oz., respectively.

Let t_1, t_2, t_3, t_4 be the tensions of the strings beginning with lowest pulley.

Then
$$t_1 = P,$$
$$t_2 = 2P + 8,$$
$$t_3 = 4P + 16 + 6,$$
$$t_4 = 8P + 32 + 12 + 4.$$

The sum of the tensions equals the weights;

therefore $15P + 78 = 1600,$

or $P = 101{\cdot}46$ oz.
$$= 6{\cdot}34 \text{ lbs.}$$

EXERCISE III.

1. If it is required to raise a weight equal to three times the power by a system of pulleys in which all the strings are attached to the weight, how many strings must there be?

2. A power of 20lbs. supports a weight of 388lbs. by means of four pulleys of equal weights, arranged according to the third system; find the weight of each pulley.

3. If in the third system there are 5 pulleys and the weights of the four movable pulleys, commencing with the lowest, are 2, 3, 4 and 5lbs., find the weight which will be sustained by a power of 12lbs.

4. What weight will be supported by a power of 12lbs., acting by means of a system of pulleys in which each cord is fastened to the weight, when there are five movable pulleys each weighing $1\frac{1}{2}$ pounds?

5. In the third system if there are eight pulleys, find the ratio that the weight of each pulley must bear to the weight supported, in order that the latter may be just supported by the weight of the pulleys alone.

6. What advantage does the third system of pulleys possess over the first or second system?

THE INCLINED PLANE.

IV.

110. Introduction. The *Inclined Plane* is a rigid plane inclined at any angle to the horizon.

The Plane is supposed to be perfectly *smooth* and perfectly *rigid*, so as to be capable of counteracting any force which acts upon it *in a direction perpendicular to its surface*. When, therefore, a body is supported on an Inclined Plane by a force directly applied to it, the case is that of a body kept at rest by three forces, its own *weight*, acting vertically downwards, the *power* applied, and the *resistance* of the plane, acting at right angles to its surface.

111. Definitions. In the annexed figure A B is called the *length* of the Plane; A C is called the *base;* and B C, which is perpendicular to A C, is called the *height* of the Plane. The angle B A C is called the *angle of inclination* of the Plane.

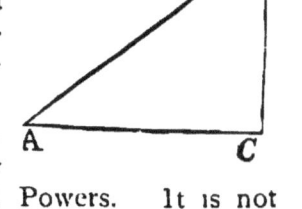

112. Remarks. It may seem that the Inclined Plane can scarcely be called one of the Mechanical Powers. It is not obvious that it can be usefully employed like the others. But it will be seen that if we have to raise a body, we may draw it up an Inclined Plane by means of a Power which is less than the Weight of the body.

SECTION I.

The Power acting parallel to the Plane.

113. *To find the ratio of the Power to the Weight, when there is equilibrium on a smooth Inclined Plane, the power acting parallel to the plane.*

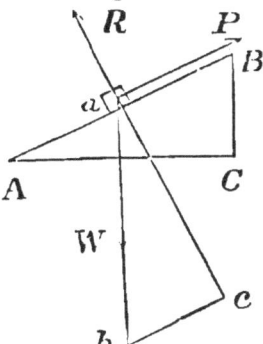

The body is kept at rest by three forces, P acting parallel to the plane, W, the weight of the body, acting vertically downwards, and R the reaction of the plane.

In the vertical line through the centre of gravity of the body, take *ab* equal AB, the length of the plane. Through *b* draw *bc* parallel to the direction of P, meeting the direction of R produced backwards.

In the triangles *abc*, ABC, we have the side *ab* equal to the side AB, the right angle *acb* equal to the right angle ACB, and the angle *bac* equal to the angle BAC, each being the complement of the angle A*ab*, therefore, the triangle *abc* is equal in every respect to the triangle ABC (Euc. 1. 26), and its sides are respectively parallel to the directions of the three forces, P, W, R, and therefore proportional to their magnitudes.

Hence, by the Triangle of Forces, we have
$$\frac{P}{W} = \frac{bc}{ab} = \frac{BC}{AB} = \frac{height}{length} = \frac{h}{l}.$$

In other words,

If in the inclined plane the power be applied parallel to the plane, the Power is to the Weight as the height of the plane is to its length.

Also, $\dfrac{P}{R} = \dfrac{bc}{ac} = \dfrac{BC}{AC} = \dfrac{height}{base}.$

Ex. I. A weight, W, is supported on an inclined plane, the inclination of which to the horizon is 30°, find the Power and the pressure on the plane.

In the preceding figure let the angle BAC = 30°. Then if AB = 2, BC will = 1, and AC = $\sqrt{3}$.

Hence $\dfrac{P}{W} = \dfrac{1}{2}$;

∴ $P = \dfrac{W}{2}.$

And $\dfrac{P}{R} = \dfrac{1}{\sqrt{3}}$;

∴ $R = P\sqrt{3}$

$= \dfrac{W}{2}\sqrt{3}.$

EXERCISE I.

1. A force of 10lbs. sustains a weight upon an inclined plane rising 2 in 5; the force acts parallel to the plane; required the weight.*

2. A railway train weighing 67,200lbs. is drawn up an inclined plane rising 1ft. in 60, by means of a rope and a stationary engine; find what number of lbs., at least, the rope should be able to support.

3. What power will be necessary to keep a weight of 50lbs. in equilibrium when the length of the plane is 6ft. and the height of the plane 1ft. 6in?

---

*An inclined plane is said to rise 2 in 5; if when its height is 2, its length is 5.

4. If a weight of 10lbs. is placed on an inclined plane whose base is 16ft. and height 12ft., and is attached by a string to an equal weight hanging over the top of the plane, find how much must be added to the weight on the plane that there may be equilibrium.

5. The length of an inclined plane is 5ft. and the height 3ft.; find into what two parts a weight of 104lbs. must be divided so that one part hanging over the top of the plane, may balance the other resting on the plane.

6. What is the vertical height of an inclined plane when the length is 2ft. 6in. and a body weighing 96lbs. is kept in equilibrium by a power of 20lbs.?

7. What is the length of an inclined plane when the vertical height is 3ft., the weight of the body 56lbs., and the power 16lbs.?

8. If a force of 40lbs., acting parallel to the length, sustains a weight of 56lbs. on an inclined plane whose base is 340ft., find the height and length of the plane.

9. What weight is that which it would require the same exertion to lift as to sustain a weight of 4lbs. upon a plane inclined at an angle of 30° to the horizon?

10. What force will sustain a weight of 56lbs., on a plane inclined at an angle of 45° to the horizon? Find also the pressure on the plane.

11. Two planes, having the same height, are placed back to back, and two weights of 7 ounces and 10 ounces respectively, connected by a string passing over the summit, are in equilibrium upon them; find the ratio of the lengths of the planes.

12. Two inclined planes, of the same height, one of which is 8ft. long, and the other 5ft., are placed so as to slope in opposite directions and so that their summits coincide. A weight of 20 ounces rests on the shorter plane, and is connected by a string passing over a pulley at the common summit of the two planes with a weight resting upon the longer plane; how great must this weight be to prevent motion?

SECTION II.

The Power acting parallel to the base.

114. *To find the ratio of the Power to the Weight, when there is equilibrium on a smooth Inclined Plane the power acting parallel to the base.*

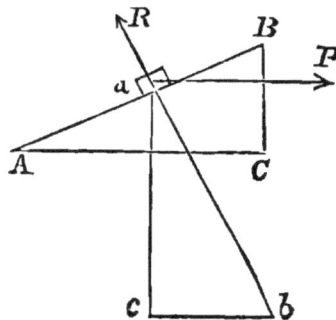

The body is kept at rest by three forces, P acting parallel to the base of the plane, W acting vertically downwards, and R the reaction of the plane.

In the vertical line through the centre of gravity of the body take ac equal AC, to which the force is parallel; through c draw ab parallel to P, meeting the direction of R produced backwards.

In the triangles acb, ACB, we have the side ac in the one, equal to the side AC in the other, the right angle acb equal to the right angle ACB, and the angle cab equal to the angle CAB, each being the complement of the angle Aac, therefore, the triangle abc is equal in every respect to the triangle ABC (Euc. 1, 26), and its sides are respectively parallel to the directions of the three forces P, W, R, and therefore proportional to their magnitudes.

Hence, by the Triangle of Forces, we have

$$\frac{P}{W} = \frac{cb}{ac} = \frac{CB}{AC} = \frac{height}{base}.$$

In other words,

If in the inclined plane, the power be applied parallel to the base, the Power is to the Weight as the height is to the base.

Also, $\dfrac{R}{W} = \dfrac{ab}{ac} = \dfrac{AB}{AC} = \dfrac{length}{base}.$

EXERCISE II.

1. What force is necessary to support a weight of 60lbs. upon an inclined plane rising 1 in 2, the force acting horizontally?

2. What weight will be supported by a horizontal force of 80lbs. upon an inclined plane rising 16 in 25?

3. A weight of 56lbs. rests upon a smooth plane inclined at an angle of 45° to the horizon. What is the smallest horizontal force required to move it up the plane?

4. What force acting horizontally will sustain a weight of 12lbs. on a plane inclined to the horizon at an angle of 60°?

5. The weight supported upon an inclined plane, whose inclination to the horizon is 45°, is $2\sqrt{2}$lbs. by a power acting parallel to the base; what is the pressure on the plane.

6. A weight of $P\sqrt{3}$lbs. is supported upon an inclined plane by a power of Plbs. acting horizontally; what is the inclination of the plane?

7. The angle of a plane is 45°; what weight can be supported by a horizontal force of 3 ounces and a force of 4 ounces parallel to the plane, both acting together?

8. A weight of 20lbs. is supported by a power of 12lbs. acting along the plane; show that if it were required to support the same weight on the same plane by a power acting horizontally, the power must be increased in the ratio of 5 to 4, while the pressure on the plane will be increased in the ratio of 25 to 16.

9. If a power which will support a weight when acting parallel to the plane be half that which will do so acting horizontally, find the inclination of the plane.

10. If R be the pressure on the plane when the power acts horizontally, and R′ when it acts parallel to the plane, show that

$$RR' = W^2.$$

THE WEDGE.

V.

115. Introduction.—The Wedge is a solid triangular prism, formed of some hard material such as steel. Its section made perpendicular to its length is generally an isosceles triangle. Its use is to separate the parts of a body; this is effected by introducing its edge between them and then applying a force to its back sufficient for the purpose.

116. *To find the ratio of the power to the resistance in an isosceles wedge.*

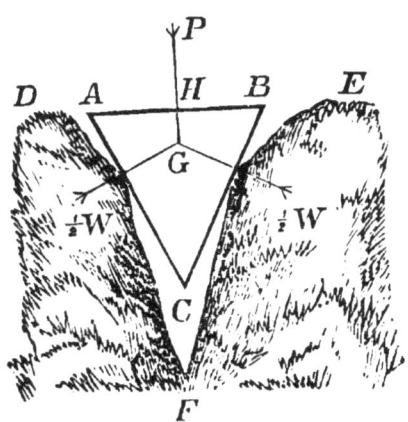

Let ABC be the section of an isosceles wedge introduced into the cleft DFE. The resistance on each side of the wedge will be the same. We may call this resistance $\frac{1}{2}W$, since the whole resistance corresponds to what we have in other machines called W. Since the wedge is smooth its action must be perpendicular to its surface; the forces $\frac{1}{2}W$, therefore, act perpendicularly to the sides of the wedge. Let the power P act at the point H, the centre of the back of the wedge. The directions of these three forces when produced will meet in a point G (art. 34). They may therefore be considered as three forces in equilibrium acting upon a point. The sides of the triangle ABC are respectively perpendicular to the directions of these forces, and therefore proportional to them (art. 32); hence we have

$$\frac{P}{\frac{1}{2}W} = \frac{AB}{BC},$$

or
$$\frac{P}{W} = \frac{\frac{1}{2}AB}{BC};$$

that is, *the power is to the total resistance as half the back of the wedge is to the side of the wedge.*

117. **Remark.**—The power usually employed with the wedge is not a pressure, but a blow, and as the proportion between a blow and a pressure cannot be well defined, the theory for calculating the power of the wedge is of little practical importance.

THE SCREW.

VI.

118. Introduction. This mechanical power is a combination of the lever and the inclined plane. It consists of a cylinder with a continuous projecting thread wound round its surface. This cylinder works in a block in which there is a groove corresponding to the projecting thread; or the cylinder may have a groove and the block a projecting thread.

The thread or groove is inclined at every point at the same angle to the axis of the cylinder.

The thread of the screw may be conceived to be generated by wrapping an inclined plane round a cylinder.

Let a piece of paper be cut in the shape of a right-angled triangle ABC; and let the side BC be placed on a pencil, or some other cylindrical surface, so as to be parallel to the axis of the cylinder. Let the paper be wrapped round the cylinder, then AB will mark out a continuous descending path, which will be the thread of the screw.

The screw is generally worked by means of a bar or arm inserted into the head of the cylinder perpendicular to its axis. One revolution of the cylinder will cause the screw to be pushed through the block or *nut* a distance equal to that between two threads of the screw, measured parallel to the axis of the cylinder.

119. *To find the ratio of the Power to the Weight in the Screw.*

Suppose the screw to be vertical and used for raising a weight W. Perpendicularly from the axis of the screw extends an arm, at the extremity of which the power P acts, P being at right angles both to this arm and to the axis of the screw. The machine is kept in equilibrium by W acting vertically downwards, P acting horizontally, and the reaction of the groove at points where the thread is in contact with it. The groove being supposed smooth, these reactions are at right angles to it.

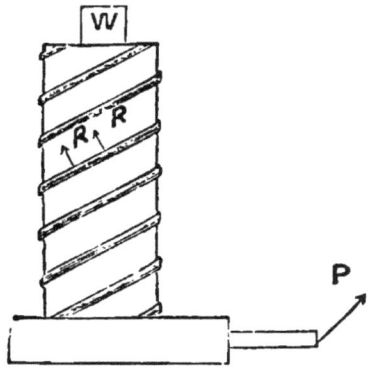

Let now a portion of the thread be conceived to be unwrapped from the cylinder; let A B, the hypothenuse of a right-angled triangle A B C, be such a portion, BC being the vertical distance between contiguous threads, and AC the circumference of the cylinder.

Let R_1 be the reaction of the groove at any point a. Construct the triangle abc equal in every respect to ABC.

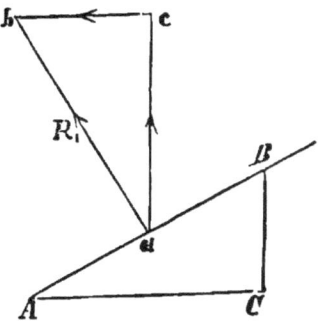

Then the vertical and horizontal components of R_1 are

$$R_1 \cdot \frac{ac}{ab}, \quad R_1 \cdot \frac{bc}{ab}; \; i.e. \; R_1 \cdot \frac{AC}{AB}, \quad R_1 \cdot \frac{BC}{AB}.$$

Suppose the other reactions R_2, R_3........similarly resolved; then, since the two sets of forces at right angles to one another form, separately, systems in equilibrium, W will be supported by the vertical, and P will be balanced by the horizontal components. Hence, we have

$$W = \frac{AC}{AB}(R_1 + R_2 + R_3 \ldots \ldots) \ldots (1).$$

Taking moments about the axis, we have

$$P.a = \frac{BC}{AB}(R_1 + R_2 + R_3 \ldots \ldots)r \ldots (2),$$

a being the length of the arm of the power, and r the radius of the cylinder.

Dividing (1) by (2) we have

$$\frac{W}{P.a} = \frac{AC}{BC.r};$$

$$\therefore \frac{W}{P} = \frac{AC}{r}.a \div BC$$

$$= \frac{2\pi a}{BC} *$$

$$= \frac{\text{circumference of Power}}{\text{distance between contiguous threads}}.$$

Ex. If the thread of the screw be inclined at an angle of 30° to a transverse section of the cylinder whose radius is 9 inches, the length of the lever which turns the screw being 4ft.; find what power will sustain 1680lbs.

If AC = the circumference of a transverse section of the screw cylinder,

then if BAC = 30°, BC will = d; and since the radius is 9 inches the circumference AC = $2\pi.9$.

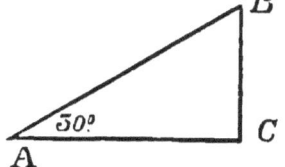

---

*If c be the circumference of a circle, then $\frac{c}{2r} = \pi$(the Greek p) = 3.14159.

Therefore $BC = \dfrac{2\pi \cdot 9}{\sqrt{3}} = 2\pi \cdot 3\sqrt{3}$.

Hence $\dfrac{P}{1680} = \dfrac{2\pi \cdot 3\sqrt{3}}{2\pi \cdot 48}$

$\qquad\qquad = \dfrac{\sqrt{3}}{16};$

therefore $P = \dfrac{1680\sqrt{3}}{16}$

$\qquad\qquad = 105\sqrt{3}$

$\qquad\qquad = 181.865 \text{ lbs.}$

EXERCISE I.

1. Describe the common screw and explain its principle. Does friction assist the power or the weight? How is the efficiency of a screw affected by increasing
   (1) the length of the arm,
   (2) the diameter of the screw,
   (3) the distance between the threads?

2. The distance between the contiguous threads of a screw is 2in., and the arm at which P acts is 20 in.; determine the ratio of P to W.

3. The circumference of the circle described by a lever working a screw is 2ft. 9in., the space between the threads is $\frac{1}{4}$ of an inch. If a force of 13lbs. be applied to the lever what will be the pressure of the screw?

4. A carriage weighing 1120lbs. is raised by a screw. The threads of the screw are $\frac{1}{8}$ of an inch apart, and the lever is 16 inches. What power is applied to work the lever?

5. The diameter of the circle described by the power is 2in., and the distance between the threads is $\frac{1}{4}$ of an inch; find the mechanical advantage.

6. The length of the arm of the power is 15in. find the distance between two consecutive threads of the screw that the mechanical advantage may be 30.

7. What must be the distance between the threads of a screw so that a power of 28lbs., applied at the extremity of a lever 25 inches long may sustain a weight of 10,000lbs. ?

8. Find the distance between the threads of a screw which is worked by an arm l, when the power applied to its extremity is an nth part of the pressure on the screw.

CHAPTER IX.

VIRTUAL VELOCITIES.

As applied to Machines.

120. **Introduction.** There is another principle from which the theory of equilibrium in machines has been derived, but which is neither so evident in itself nor capable of such an easy demonstration as the principle of the composition and resolution of forces, and the principle of moments. It has, however, the advantage of being much more conveniently applied to questions of equilibrium of complicated machinery. It is called the Principle of Virtual Velocities.

121. **Def. of Virtual Velocity.** *If a machine is in equilibrium under the action of a power P and a weight W, and we suppose it to receive any displacement, consistent with the connection of its various parts, then the spaces described by P and W, estimated in their respective directions, are called the Virtual Velocities of the power and weight.*

122. **Remarks.** The word *virtual* is used to indicate that the motion referred to, does not really take place, but is only *supposed* to take place. The word *velocities* is used because we may conceive the points of application of P and W to move into their new positions in the same time, and then the lengths of their paths described will be proportional to their velocities.

123. *Principle of Virtual Velocities.* The Principle of Virtual Velocities asserts that in any machine the Power multiplied by its virtual velocity is equal to the Weight multiplied by its virtual velocity.

This important principle is only a particular case of a far more general one which we cannot here examine; nor can we even give a general proof of the principle as stated above. All that we shall be able to do is to prove that the principle holds good in those cases of equilibrium which we have already established by other methods.

124. *To apply the Principle of Virtual Velocities to find the ratio of the Power to the Weight in the lever.*

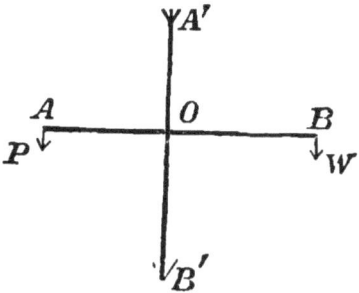

Suppose the lever to be a straight lever AB, having its arms $AO = a$, and $BO = b$, and to be acted on by forces P and W perpendicular to the arms.

Let the lever AB be turned vertically through a right angle into the position A'B', P and W remaining constant in direction; then,

the virtual velocity of P is OA' or a,
and "   "   W is OB' or b.

By the Principle of Virtual Velocities we have
$$P \times a = W \times b,$$
the condition of equilibrium established by the principle of moments.

In a similar manner the principle may be applied to find the ratio of the power to the weight in a lever of second or third order. Hence, the Principle of Virtual Velocities holds in the case of the Lever.

125. **The Wheel and Axle.** Conceive the machine to make one complete turn; then,

the virtual velocity of P = circumference of wheel,
" " W = " axle;

and the Principle of Virtual Velocities asserts that
P × circumference of wheel = W × circumference of axle.

Divide both sides of this equation by 2π, and

P × radius of wheel = W × radius of axle,

the result obtained in Article 99, and hence the Principle holds in the case of the Wheel and Axle.

126. **The Pulley.** In applying the Principle of Virtual Velocities to pulleys, we suppose the weight to be raised through a small space, which space will be its virtual velocity, and the corresponding space through which the point of application of P must be moved will be the virtual velocity of P.

127. **The Single Movable Pulley.** In the figure Article 104, suppose W raised through 1 inch, the string on either side of the pulley B will be shortened by the same quantity; consequently P will descend through 2 inches; hence,

the virtual velocity of P = 2 inches,
" " W = 1 inch;

and the Principle of Virtual Velocities asserts that

P × 2 = W × 1,

the condition found in Article 104.

128. **The First System of Pulleys.** In the figure Article 105, let W be raised through 1 inch, then the pulley C rises through 1 inch, the pulley B, through 2 inches, the pulley A through 2 × 2, or 2^2 inches, and

so on; hence, if there were n movable pulleys, the nth pulley will rise through $2n-1$ inches, and P will descend through 2^n inches; hence,

$$W\text{'s virtual velocity} = 1 \text{ in.,}$$
$$P\text{'s} \quad " \quad " \quad = 2^n \text{ in.;}$$

and the Principle of Virtual Velocities asserts that

$$P \times 2^n = W \times 1,$$

which is the result obtained in Article 105, and, therefore, the Principle holds in the case of first system of pulleys.

129. The Second System of Pulleys. In the figure Article 107, let W be raised through 1 inch, then if there were n strings between the two blocks, each of them will be shortened by 1 inch; consequently P will descend through n inches: hence,

$$W\text{'s virtual velocity} = 1 \text{ in.,}$$
$$P\text{'s} \quad " \quad " \quad = n \text{ in.;}$$

and by the Principle of Virtual Velocities we have

$$P \times n = W \times 1,$$

the condition obtained in Article 107. The Principle is, therefore, true in the second system of pulleys.

130. The Third System of Pulleys. In the figure Article 108, let W be raised through 1 inch; then the highest movable pulley descends through 1 inch. The next pulley descends through 2 inches in consequence of the descent of the pulley above it, and through 1 inch in addition, in consequence of the ascent of the weight; hence on the whole its descent will be $(2 + 1)$ inches. The next pulley descends through twice this, in consequence of the descent of the pulley above it, and through 1 inch besides in consequence of the ascent of the weight; therefore its whole descent will be $\{2(2+1)+1\}$ inches, that is $(2^2 + 2 + 1)$ inches. In a similar manner it may be shown that P descends through $(2^3 + 2^2 + 2 + 1)$ inches, hence,

the virtual velocity of $P = (2^3 + 2^2 + 2 + 1)$ inches
$$= (2^4 - 1) \text{ inches,}$$
and " $W = 1$ inch;

and by the Principle of Virtual Velocities we have
$$P \times (2^4 - 1) = W \times 1,$$
the result obtained in Article 105.

131. The Inclined Plane. In the figure Article 113, let W be at A and suppose it drawn up to B; then P's displacement in its own direction is the length of the Plane. Now W's direction is vertical, and its displacement in that direction is evidently the height of the plane; hence

$$\text{P's virtual velocity} = AB,$$
$$\text{W's} \qquad \text{''} \qquad = BC;$$

and the Principle of Virtual Velocities assert that
$$P \times AB = W \times AC,$$
the condition obtained in Article 113; hence the Principle holds in the case of the inclined plane, when the power acts parallel to the plane. In a similar manner it may be shown that the Principle holds when the power acts parallel to the base.

132. The Screw. If the arm on which P acts be made to describe a complete revolution, the weight will be raised or depressed through a space equal to the vertical distance between two threads of the screw; hence,

P's virtual velocity = circumfer. of circle described by P,
W's   ''   = vertical dist. between two threads;
and the Principle of Virtual Velocities asserts that
P × circumference of circle described by P = W × distance between two threads, which is the result obtained in Article 119.

The Principle, therefore, holds in case of the screw.

133. Remarks. We have thus shown that the Principle of Virtual Velocities holds good in all the simple machines, since it always leads to correct results; we may, therefore, safely conclude that it holds good in any combination of these machines. We have, therefore, a principle of great beauty, and very easy of application which brings all cases of equilibrium in machines under one general law, and enables us to find the ratio of the power to the weight in machines of the most complicated construction.

If P be always supposed to describe the same space in the same time, it will easily be seen that "what is gained in power in any of the above machines is lost in time." This is sometimes stated as a mechanical principle.

EXERCISE I.

1. Explain the term *virtual velocity*. State the Principle of Virtual Velocities, and show that it holds good in a lever of the third order.

2. State the relation between two weights which balance one another on a given wheel and axle.

Prove that the Principle of Virtual Velocities is true in this instance.

3. Show that the Principle of Virtual Velocities holds in the case of a body in equilibrium on a smooth inclined plane, the power acting parallel to the base of the plane.

4. When a power of 10lbs. is applied to lift a weight through 2 inches, the power descends through 3 inches; find the weight.

5. With a wheel and axle a power of 14lbs. balances a weight of 2240lbs.; if the power descended through 80 inches, through what height would the weight be raised?

6. In the wheel and axle the radius of the wheel is 15 times that of the axle, and when the weight is raised through a certain height it is found that the point of application of the power has moved over 7ft. more than the weight; find the height through which the weight has been raised.

7. A wheel and axle is used to raise a bucket from a well. The radius of the wheel is 15in., and while it makes 7 revolutions, the bucket, which weighs 30lbs., rises $5\frac{1}{2}$ft. What is the smallest force that can be employed to turn the wheel?

8. In the first system of pulleys the distance of the highest pulley from the fixed end of the string which passes round it is 16ft.; find the greatest height through which the weight can be raised, there being four pulleys.

9. In the second system of pulleys, how much cord would be required for a man to raise himself 30ft., there being three pulleys in the lower block?

10. Employ the Principle of Virtual Velocities to find the weight which 1lb. will support by means of two blocks of pulleys, of which the upper one is fixed, and the lower one, weighing 1lb., is movable, each block containing 4 pulleys, and the portions of string between the blocks being all vertical.

11. A weight of 253lbs. is supported on an inclined plane rising 8 ft. in a length of 137ft. by two equal forces one being horizontal and the other parallel to the plane; find the forces.

12. Two weights hang over a pulley fixed to the summit of a smooth inclined plane, on which one weight is supported, and for every 3 inches that one is made to descend, the other rises 2 inches; find the ratio of the weights and the length of the plane, the height being 1ft. 6in.

13. If a power of 1lb describes a revolution of 3 feet whilst the screw moves through $\frac{1}{4}$in., what pressure will be produced?

CHAPTER X.

FRICTION.

134. **Def. of Friction.** In the preceding investigations we have supposed that all surfaces are smooth. Practically this is not the case. When we attempt to make the surface of a body move upon that of another, with which it is kept in contact by pressure, there is, in general, a resistance to motion; this resistance is called *friction*.

135. **Direction in which Friction acts.** Friction at any point of a surface acts in the direction exactly contrary to that in which motion would occur if the surface were smooth. Thus, if a body on an inclined plane, under the action of any force, is on the point of *ascending*, the force of friction acts downwards; but if the body is on the point of *descending*, the action of friction is upwards. Hence a given weight may be sustained by a less power than is required independently of friction, but requires a greater power to move it.

136. **Laws of Friction.** The following laws, which have been established by experiment, apply to a body which is on the point of moving on the surface of another body, and is only prevented from doing so by friction.

I. *The friction between two surfaces of the same kind is in direct proportion to the pressure between them.*

II. *The amount of friction is independent of the extent of the surface in contact, so long as the pressure remains the same.*

137. Coefficient of Friction. Let R be the perpendicular pressure between two surfaces, F the friction, then by the first Law, the ratio of F to R is a constant number for two given substances. This number is called the *coefficient of friction* and is denoted by μ (the letter m of the Greek alphabet), that is

$$\frac{F}{R} = \mu, \text{ and therefore } F = \mu R.$$

138. *To find the coefficient of friction between two given substances.*

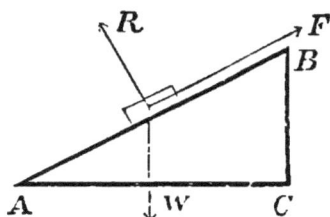

Let one of the substances form an inclined plane AB movable about the point A, and let a block of the other substance, having a plane base, be placed upon it. Let the plane AB be elevated till the body is on the point of sliding down it. The body is then supported on the plane by the three forces F, W, and R ; hence we have

$$\frac{F}{R} = \frac{BC}{AC} = \mu \text{ (art. 137)}.$$

The value of this coefficient has been found by experiment for various substances; thus, for metals on metals it is about .17 ; for wood on wood it has been found to be about .33.

The angle BAC is called *the angle of repose*, and sometimes *the limiting angle of friction*.

FRICTION.

EXAMPLES.

1. A body is just on the point of sliding on a rough plane that rises 3 in a length of 5; find the coefficient of friction.

In the preceding figure let $AB=5$ and $BC=3$, then $AC=4$.

The body is kept at rest on the plane by the forces F, W, and R; hence we have

$$\frac{F}{R} = \frac{BC}{AC} = \frac{3}{4} = \mu.$$

The coefficient of friction is, therefore, $\frac{3}{4}$.

2. What force, acting parallel to the plane, will be required to draw a square block of timber up a plane inclined to the horizon at an angle of 60°, the coefficient of friction being .63, and the weight of the block 5 cwt.?

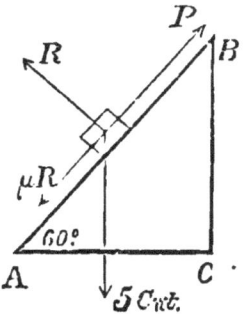

Since the body is on the point of moving up the the plane, friction acts downwards (art. 135).

The body is kept in equilibrium on the plane by the forces P, R, friction which is equal to μR, and the weight, 5 cwt.; of these forces P and μR are equivalent to the single force $P - \mu R$ acting up the plane.

Since the angle BAC = 60°, the sides AB, BC, AC, of the right-angled triangle ABC, are in the ratio of 2, 1 and $\sqrt{3}$; hence we have

$$\frac{P-\mu R}{5} = \frac{\sqrt{3}}{2} \quad \ldots\ldots\ldots\ldots\ldots\ldots(1),$$

and
$$\frac{R}{5} = \frac{1}{2} \quad \ldots\ldots\ldots\ldots\ldots\ldots(2).$$

Substitute the value of R from (2) in (1), and for μ substitute its value .63, and we have

$$P - .63 \times \frac{5}{2} = \frac{5\sqrt{3}}{2};$$

$$\therefore P = \frac{5}{2}(\sqrt{3} + \cdot 63)$$

$$= \frac{5}{2}(1.732 + .63)$$

$$= 5.9 \text{ cwt.}$$

EXERCISE I.

1. What is meant by friction, and by the angle of repose?

2. What is meant by the coefficient of friction? How is it related to the limiting angle of resistance?

3. What are the laws of friction? How may the co-efficient of friction for different substances be determined?

4. What are the forces acting on a body, which stands at rest on the side of a hill?

5. A ladder AB has one end A on a rough horizontal road, and the other end B, against a rough vertical wall; what are the forces acting on it?

6. If the coefficient of friction is $\frac{1}{\sqrt{3}}$, show that the inclination of plane is 30°.

7. A body rests without support on a plane inclined to the horizon at an angle of 45°; what is the coefficient of friction between the body and the plane?

8. A plane rises 5 in 13 ; find the greatest weight which can be sustained upon it by a power of 20lbs. acting along the plane, the coefficient of friction being $\frac{1}{4}$.

9. Find the least force which will sustain a weight of 10lbs. upon a plane rising 7 in 25 when the force acts along the plane, and the coefficient of friction is $\frac{1}{4}$.

10 Find the least force which will draw a weight of 10lbs. up a plane rising 11 in 122, when the force acts along the plane, and the coefficient of friction is $\frac{5}{12}$.

11. A body placed on a horizontal plane requires a horizontal force equal to its own weight to overcome the friction ; supposing the plane gradually tilted up, find at what angle the body will begin to slide.

Ex. 3. A ladder rests against a vertical wall to which it is inclined at an angle of 45°; the centre of gravity of the ladder is $\frac{1}{3}$ the length from the foot. The coefficient of friction for the ladder and plane is $\frac{1}{4}$, and and for the ladder and wall $\frac{1}{4}$. If a man whose weight is half the weight of the ladder ascends it, find to what height he will go before the ladder begins to slide.

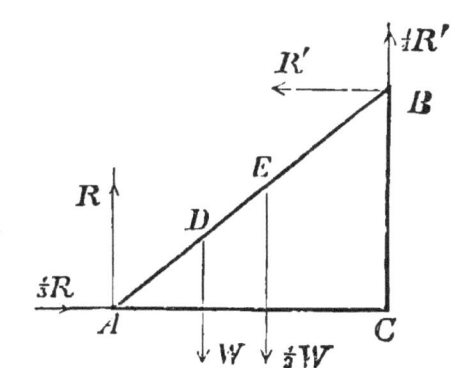

Let AB, the ladder, $= 3a$ ft. in length.

„ $AE = x$, be the height to which the man has ascended before the ladder begins to slide.

The weight of the ladder, W, acts at D, its centre of gravity, AD being equal to a.

When the man has ascended to E, the ladder is kept in equilibrium by the forces as indicated in the figure.

Since the forces are in equilibrium, we have

$$R' = \frac{1}{3} R,$$

and
$$R + \tfrac{1}{4}R' = \frac{W}{2} + W;$$

therefore
$$R' = \frac{6W}{13} \quad \dots\dots\dots\dots\dots(1).$$

Take moments about A, and we have

$$\frac{W}{2} \times \frac{x}{\sqrt{2}} + W \times \frac{a}{\sqrt{2}} = R' \times \frac{3a}{\sqrt{2}} + \frac{1}{4}R' \times \frac{3a}{\sqrt{2}},$$

or
$$\frac{Wx}{2} + Wa = R' \times \frac{15a}{4}$$

$$= \frac{6W}{13} \times \frac{15a}{4}, \text{ from (1)};$$

$$\therefore x = \frac{19}{13} a = \frac{19}{39}.3a = \frac{19}{39} \text{ of}$$

the length of the ladder.

EXERCISE II.

1. A uniform ladder resting at an angle of 45° between a smooth vertical wall, and a rough horizontal plane, is just supported by the friction of the ground; find the coefficient of friction.

2. A ladder inclined at an angle of 60° to the horizon rests between a rough pavement and a smooth wall of a house. Show that if the ladder begins to slide when a man has ascended so that his centre of gravity is half way up, then the coefficient of friction between the foot of the ladder and the pavement is $\tfrac{1}{6}\sqrt{3}$.

3. A ladder 10 feet long, rests with one end against a smooth vertical wall, and the other on the ground, the coefficient of friction being .5; find how high a man,

whose weight is three times that of the ladder may ascend before the ladder begins to slip, the foot of the ladder being 6 feet from the wall.

4. A beam 21ft. long rests with its ends upon a horizontal and a vertical plane, respectively; find how far the lower end may be drawn out from the wall before the beam will slip; the coefficients of friction being $\frac{1}{4}$ and $\frac{1}{6}$, respectively.

5. A uniform ladder is placed between a rough horizontal plane and a rough vertical wall, at an angle of 45°, the coefficient of friction between the ladder and the ground being $\frac{4}{8}$; a man whose weight is half that of the ladder ascends; find what the coefficient of friction must be between the ladder and the wall, that when the man reaches the top of the ladder it may just begin to slip.

6. A ladder which is divided by its centre of gravity into segments of 10ft. and 20ft., is placed against the side of a house at an angle of 30°; find the highest round upon which a weight of 4cwt. can be suspended; the weight of the ladder being 3cwt., and the coefficient of friction between the ladder and the wall, and also between the ladder and the horizontal pavement which supports its heavier end being $\frac{1}{\sqrt{3}}$.

CHAPTER XI.

The application of Similar Triangles to the solution of statical problems.

Introduction. Some of the questions set to candidates for first class certificates require a knowledge of a few propositions in Euclid B. VI. The following are the most important:—

If a straight line be drawn parallel to one of the sides of a triangle, it shall cut the other sides, or those sides produced proportionally; and if the sides, or the sides produced, be cut proportionally, the straight line which joins the points of section, shall be parallel to the remaining side of the triangle. Euc. B. vi. 2.

Similar Triangles. Two triangles are said to be *similar* when they are equiangular.

The sides about the equal angles of similar triangles are proportionals; and those which are opposite to the equal angles are antecedents or consequents of the ratios. (Euc. B. vi. 4).

EXAMPLES.

1. A uniform beam AB, 12 feet long and weighing 40lbs., rests with one end A at the bottom of a vertical wall, and a point C, 2ft. from the other end, is connected by a string CD to a point D, 8ft. above A; find the reaction at A, and the tension of the string.

The beam is supported by three forces—

1. T, the tension of the horizontal string along CD.
2. W acting vertically through G, 6ft. from A, its direction meeting CD in E.
3. The reaction R at A along AE.

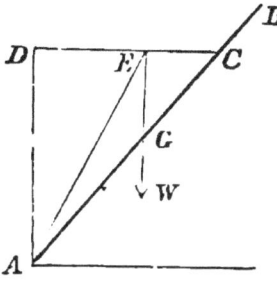

Since G is the middle point of AB, AG=6, and GC=4; DC is, therefore, cut in the point E in the ratio of 6 to 4 (Euc. vi. 2).

Hence, $DE = \dfrac{6}{10} \times 6 = 3.6$ feet.

And $DE^2 + AD^2 = AE^2$; $\therefore AE = 8.77$.

From the triangle of forces, we have

$$\dfrac{T}{W} = \dfrac{DE}{AD},$$

or $\dfrac{T}{40} = \dfrac{3.6}{8}$;

$\therefore T = 18$lbs.

Similarly $R = 43.8$lbs.

2. A rod 8 feet long, the weight of which has not to be considered, is placed across a smooth horizontal rail, and rests with one end against a smooth vertical wall, the distance of which from the rail is 1 foot; the other end of the rod bears a weight of 12lbs. Find the position of equilibrium and the pressure on the rail.

The three forces acting on the rod AB, are the weight 12lbs., at B vertical, the reaction of the wall at A horizontal, and the pressure on the rail C perpendicular to AB. Since the directions of the first two meet in E, the third must be along EC. Hence the position of equilibrium is such that EC is perpendicular to AB.

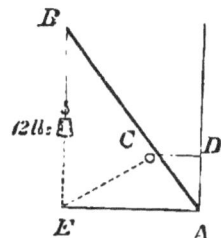

Draw CD horizontal. Then the triangles ACD, ACE, ABE, are similar; hence we have (Euc. vi. 4)

$$\frac{CD}{CA} = \frac{CA}{AE} = \frac{AE}{AB};$$

therefore $CA^2 = AE.CD$
$= AE$, since $CD = 1$.

And $AE^2 = CA.AB$
$= 8CA$, since $AB = 8$.

therefore $CA = 2$,
and $AE = 4$.

Therefore the angle $CAE = 60°$ (art. 28).

Again the triangle EAB has its sides perpendicular to the directions of the three forces acting on the rod, and therefore proportional to them (art, 32). Hence if P be the pressure on the rail, we have

$$\frac{P}{12} = \frac{8}{4};$$

$$\therefore P = 24 \text{lbs}.$$

3. A weight W, of 34lbs, suspended from a string MD, which is attached to a fixed point M, rests at D on an inclined plane CA, whose base AB is 39 inches; and its vertical height CB, 26 inches. The position of D is such that BD is at right angles to CA; and the point M is in BC produced, CM being 8 inches. Prove that the tension of the string MD is 20lbs.; and find the reaction of the plane on the weight.

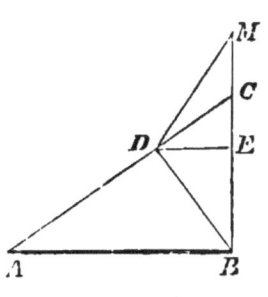

First Class Exam., July, 1873.

The weight W, is kept at rest by the three forces —
1. The tension of the string T, acting along DM.
2. The weight W, acting vertically downwards.
3. The reaction R, at right angles to the plane and therefore in BD produced.

Hence the triangle BDM has its sides parallel to the three forces and is, therefore, proportional to them.

SIMILAR TRIANGLES.

Since $AB = 39$, and $BC = 26$, $\therefore AC = 13\sqrt{13}$.

Since the triangle BCD is similar to the triangle ACB, we have

$$\frac{DC}{CB} = \frac{CB}{AC},$$

or

$$\frac{DC}{26} = \frac{26}{13\sqrt{13}};$$

$$\therefore DC = 4\sqrt{13}.$$

Also

$$\frac{BD}{BC} = \frac{AB}{AC},$$

or

$$\frac{BD}{26} = \frac{39}{13\sqrt{13}};$$

$$\therefore BD = 6\sqrt{13}.$$

Draw DE parallel to AB. The triangle CDE is similar to the triangle CAB; hence we have

$$\frac{DE}{DC} = \frac{AB}{AC};$$

$$\therefore DE = 12.$$

Also

$$\frac{CE}{ED} = \frac{CB}{AB};$$

$$\therefore CE = 8.$$

And

$$DM^2 = DE^2 + EM^2$$

$$= 12^2 + 16^2;$$

$$\therefore DM = 20.$$

From the triangle of forces, we have

$$\frac{T}{W} = \frac{DM}{MB},$$

or

$$\frac{T}{34} = \frac{20}{34};$$

$$\therefore T = 20.$$

And

$$\frac{R}{W} = \frac{BD}{BM},$$

or

$$\frac{R}{34} = \frac{6\sqrt{13}}{34};$$

$$\therefore R = 6\sqrt{13} \text{ lbs.}$$

We might have obtained the preceding results without using similar triangles.

Find AC as before. Then
AC × BD = AB × CB, since each product is double triangle ABC;

$\therefore$ BD = $6\sqrt{13}$.

Again, in the triangle BCD, we have
$$DC^2 = BC^2 - BD^2;$$

$\therefore$ DC = $4\sqrt{13}$.

And DE × BC = BD × DC, each product being double triangle BDC;

$\therefore$ DE = 12.

Also $EC^2 = DC^2 - DE^2$;

$\therefore$ EC = 8.

And DM = 20, as before.

EXERCISE I.

1. If two forces P and W sustain each other on the arms of a bent lever PCW, whose fulcrum is C, and act in directions PA, WA, which form the sides of an isosceles triangle PAW; show that if AC be joined, and produced to meet PW in F,

$$P : W :: FW : FP.$$

2. If a weight be suspended from one extremity of a rod movable about the other extremity A, which remains fixed, and a string of given length be attached to any point B in the rod, and also to a fixed point C above A in the same vertical line with it, then the tension of the string varies inversely as the distance AB.

3. Three forces act on a particle, their directions being parallel to the three perpendiculars drawn from the angles of a triangle to the opposite sides, and their magnitudes inversely proportional to these perpendiculars; show that the three forces are in equilibrium.

4. A uniform beam AB, whose weight is W, and length 6 feet, rests on a vertical prop CD equal to 3 feet; the other end A is on the horizontal plane AD, and is prevented from sliding by a string DA equal to 4 feet; find the tension of this string.

5. Find the centre of gravity of a conical shell contained between two right circular conical surfaces, having the same axis, the outer diameter of the shell being 8 inches, the inner diameter 6 inches, and the height of the whole 12 inches.

6. A uniform beam AB whose weight is 50lbs., rests with one end B on the top of a vertical wall, the height of which is half the length of the beam, the other end A is on the horizontal plane AD, and is prevented from sliding by a string DA equal to $\frac{2}{3}$ of AB; find the tension of the string.

7. A 9-pounder gun, weighing 13cwt. 2qrs. is suspended at the extremity B of a spar AB, which is supported by a stay BC, fixed at C; find the tension of the stay and the pressure on the spar when AB = 13 feet, AC = 10 feet, and AD, the horizontal distance of the gun from A, = 5 feet.

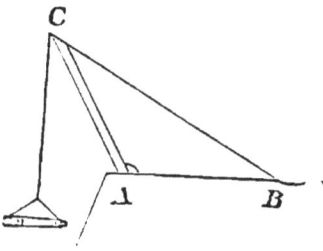

8. A gun, to be lowered over the parapet of a fortress, is suspended from the extremity C of a spar AC, 10 feet long, held in its position by a stay BC 15 feet in length, the distance from A to B being 10 feet; find the tension of BC and the thrust upon AC, when AB is horizontal, and the weight of the gun 33cwt.

9. Two weights, connected by a string passing over the common vertex of two inclined planes, balance each other; if they are set in motion, show that their centre of gravity will move in a horizontal straight line.

10. A ladder 10 feet long, rests with one end against a smooth vertical wall, and the other on the ground, the coefficient of friction being .5; find how high a man, whose weight is three times that of the ladder, may ascend before the ladder begins to slip, the foot of the ladder being 6 feet from the wall.

CHAPTER XII.

Examination Papers in Statics set to Candidates for First and Second Class Certificates 1871.

SPECIAL EXAMINATION FOR COUNTY INSPECTORS.

1. State the principle of the Parallelogram of Forces. What is the magnitude of the resultant of five forces acting on a particle, and represented respectively in magnitude and direction by the sides of a regular pentagon taken in order?

2. Draw a system of pulleys in which the relation between the power and the weight is expressed by the formula $W = P \times 2^n$, n being the number of movable pulleys. (2) Should the weight of the pulleys be taken into account, what is the relation between W and P, when there are four movable pulleys in the system, and the weight of each is $\dfrac{W}{15}$?

3. Enunciate the principle of Virtual Velocities; and assuming the principle, apply it to find the relation between the power and the weight in the screw.

4. On a plane inclined to the horizon at an angle of $30°$, a weight W is supported by a power P, acting parallel to the plane. Show that a power equal to 2P, acting parallel to the base, would be required to sustain W on a plane inclined to the horizon at an angle of $45°$.

5. One end of a uniform beam whose weight is 100lbs., rests on a horizontal plane, and in a vertical plane passing through the beam. Find the reaction of the ground against the beam, and the amount of friction.

SECOND CLASS CERTIFICATES.

JULY 1871.

1. In a straight lever CBA, a weight W of 10lb. acts at B, a distance of 5 feet from the fulcrum A; and the Power P is applied at C, on the same side of the fulcrum as W, but in an upward direction. If AC be $12\frac{1}{2}$ feet, what is P? And what is the pressure on the fulcrum?

2. Assuming the principle of Virtual Velocities, deduce therefrom the relation that must subsist between the Power and the Weight in the lever, in order that there may be equilibrium.

3. At points A, B, C, which are in the same straight line, weights of 8lbs., 12lbs., and 20lbs., respectively, are placed. If AB be $12\frac{1}{2}$ feet, and BC be 5 feet, it is required to find the centre of gravity of the three weights.

4. (*a*) In a system of pulleys where each pulley hangs by a separate string, what power will sustain a weight of 104lbs., there being two movable pulleys in the system?

(*b*) If the weight of each of the pulleys be $1\frac{1}{2}$lbs., by what power will the weight of 104lbs. be sustained, the weight of the pulleys being taken into account?

5. A weight of 80lbs. is sustained on an inclined plane by a power of 60lbs. acting parallel to the base. How many feet does the plane rise in the hundred?

6. (*a*) One end of a uniform beam rests on the ground, the other being supported in the hands of a man who exerts a pressure which is at right angles to the length of the beam, and in a vertical plane passing through the beam. Mention the different forces by which the beam is kept at rest.

(*b*) If the weight of the beam be 200lbs., and its inclination to the ground 60°, what force does the man exert?

FIRST CLASS CERTIFICATES.

July 1871.

1. (*a*) Enunciate the principle of Virtual Velocities.

(*b*) Assuming the principle, deduce the relation that must subsist between the Power and the Weight in the inclined plane, in order that there may be equilibrium; pointing out *precisely* the quantities which represent the virtual velocities of the Power and the Weight respectively; the Power being supposed to act parallel to the plane.

2. In a straight lever, the lengths CA, CB, measured to the left of the fulcrum C, are 6 and 10 feet respectively; and CD, CE, measured to the right are 5 and 8 feet respectively. Mention any weights which, suspended from A, B, D, E, respectively, will keep the lever at rest.

3. A balance has one of its arms 6 inches long and the other $6\frac{1}{2}$ inches. A package of tea is weighed out to a purchaser in the one scale against the weight of 2 lbs. in the second scale; and another package is weighed out to him in the second scale against a weight of 2 lbs. in the first scale. How much tea does he get altogether?

4. A weight W is just sustained on an inclined plane by two forces, one of 9 lbs. acting parallel to the base and toward the plane; the other of 21 lbs. acting parallel to the plane, and in an upward direction. The base of plane being 40 feet, and its perpendicular height 30 feet, what is W?

SECOND CLASS CERTIFICATES.

December 1871.

1. Describe the Wheel and Axle, stating the relation between the power and the weight necessary in order to equilibrium.

2. Describe the Screw, stating the relations between the power and the weight, necessary in order to equilibrium.

3. Show that the principle of Virtual Velocities holds in the Screw.

4. A body, situated at A, one of the angular points of the square ABCD, whose diagonal is AC, is acted on by three forces, represented in magnitude and dirction by the lines AB, AC, AD, respectively. Draw a line representing in magnitude and direction, a fourth force which, in connection with the three above-mentioned, would keep the body at rest.

5. A uniform straight lever ABC, whose fulcrum is B, weighs 10lbs. The arm AB is 12 feet; the arm BC, 8 feet. Weights of 8 and 12lbs. are suspended from A and C respectively. Find the additional weight which must be suspended from C, in order to keep the lever at rest, the weight of the lever being taken into account.

FIRST CLASS CERTIFICATES.

December 1871.

1. In a system of pulleys, each of which hangs by a separate string, apply the principle of Virtual Velocities to determine the relation between the power and the weight when there is equilibrium.

2. In the previous question should each pulley weigh 2lbs., and should the weight of the pulleys be taken into account, what power would be required to sustain a weight of 40lbs., there being two movable pulleys in the system?

3. ABCD is a square; and, at a point E, in the diagonal AC, is placed a particle to which are attached strings that pass over smooth pulleys at the corners, A, B, C, D, of the table, and support weights represented in magnitude by the lines CE, DE, AE, BE, respectively. Show that the particle will be kept at rest.

NORMAL SCHOOL

1871.

1. A uniform straight lever ABC, whose fulcrum is B, has the arm AB 10 feet, and BC 8 feet. A weight of 20lbs. hangs from the middle point of the lever, and another of 30lbs. from the extremity A. What weight must be suspended from C to produce equilibrium?

2. Describe the Differential Wheel and Axle The radii of the axles are $4\frac{1}{2}$ and 5 inches, respectively. What power will susain a weight of 200olbs.?

3. What weight will be supported on an inclined plane by a power of 6olbs. acting parallel to the base; the height of the plane being 7 feet, and the base 17 feet?

4. Define the centre of gravity. How can it be shown that every system of particles has a centre of gravity? Where is the centre of gravity of two heavy particles? In the straight line AD take AB = 3 ft., BC = 1 ft., CD = 1 ft.; and place at B, C, D, weights of 3lbs., 2lbs, 1lb., respectively. Then if d be the distance of the centre of gravity of these three weights from A prove that

$$(1 + 2 + 3)\,d = (3^2 - 0^2) + (3^2 - 1^2) + (3^2 - 2^2).$$

5. A uniform beam AB, 20 feet long, and weighing 900lbs., rests in a horizontal position on equal upright props CA and DB. It is loaded at a point E, 5 feet from A, with a weight of 100lbs.; find the pressure on each prop.

SECOND CLASS CERTIFICATES.

JULY 1872.

1. Describe the two kinds of levers, and give two examples of each kind.

2. In a bent lever, AFB, whose fulcrum is F, the arms AF and FB are straight, and inclined to one another at an angle AFB, which is such, that the perpendicular let fall from A on BF produced is equal to two-fifths of AF. Also FB = 2FA. When FB is horizontal, find the force, which, acting vertically at A, will balance a force of 100lbs. acting vertically at B.

3. Describe the Screw.

4. The base, AC, of an inclined plane, AB, is 8ft., and the vertical height BC, 6 feet. A weight W, of 100lbs. is sustained on AB by a force P, acting parallel to AB. Find the reaction of the plane against W (or the pressure of W on the plane in a direction perpendicular to the plane).

5. The triangle ABC has the angle ACB a right angle; AC = 3 feet, and BC = 8 feet. From A let the straight line AD be drawn to D, the middle point of BC; and suppose that a particle at A is acted on by three forces represented in magnitude and direction by the lines AB, AD, AC respectively. Find the length of the line which represents the sum of the resolved parts in a direction perpendicular to AC; and deduce the magnitude of the resultant of the three forces.

6. In sliding friction what is the relation between friction and pressure? How does friction depend on the extent of surface in contact, and on the velocity of motion?

FIRST CLASS CETIFICATES.
July 1872.

1. A straight lever ACB, whose fulcrum is C, is kept at rest by two forces P and Q, the former acting at A in the direction AP, at the latter at B in the direction BQ. The lines PA and QB produced meet in a point D. Show that, if the forces P and Q be represented in magnitude by the lines DA and DB respectively, CA is equal to CB.

2. On the same base AB, and on the same side of it, are described the equilateral triangle ABC, and the isosceles triangle ABD, which is right-angled at B. A particle at A is acted on by two forces, one in the direction AC, the other in the direction DA (not AD); and these forces are represented in magnitude by lines which are 2 feet and $\sqrt{2}$ feet in length, respectively. Prove that their resultant is at right angles to AB.

SECOND CLASS CERTIFICATES.
July 1873.

1. The mechanical advantages of a machine, on which a weight W is sustained by a power P, being understood to be the ratio $\dfrac{W}{P}$; and the machine being said to gain, or to lose, advantage, according as this

ratio is greater or less than unity; describe two kinds of straight levers, the one of which always gains, and the other always loses advantage.

When mechanical advantage is lost, does the machine serve any good purpose? If so, what? Illustrate by a familiar example.

2. When a door is turned round its hinges by a force applied horizontally at the handle, what is the resistance which the force applied overcomes? If the weight of the door be increased, is the force required to turn the door increased? If so, how does the increase of a *vertical* force render necessary the increase of a *horizontal* force?

3. "A number of pulleys are attached to the same block, which supports a weight, and the same string passes round all the pulleys." Draw a diagram representing such a system, with three pulleys in the block; and state (without proof) the mechanical advantage.

4. Enunciate the principle of Virtual Velocities; and show how it holds good in the system of pulleys referred to in the previous question.

5. What is friction?

Inquire whether friction acts to the advantage or to the disadvantage of the Power, *first*, when the Power is on the point of raising the Weight; and *secondly*, when the Power is just preventing the Weight from descending. What is the coefficient of friction?

FIRST CLASS CERTIFICATES.

JULY 1873.

1. A particle at E, a point on a square surface ABCD, is kept at rest by four forces, which are represented in *direction* by the lines EA, EB, EC, ED, respectively; and in *magnitude* by EA, 2EB, 3EC, 4ED, respectively. Prove that the point E is equally distant from one pair of the opposite sides of the square; and that its distances from the other pair of opposite sides are in the proportion of 7 to 3.

2. Let BA, BC, represent a double inclined plane; the length of BA being 25 feet, and that of BC 40 feet; AC being horizontal. A weight P, resting on AB, is

balanced by a weight W, resting on BC; P and W being connected by a string, which passes over a fixed pulley at B, and the two portions of which are parallel to AB and BC respectively. If W be 100lbs., find P. Inquire whether the tension of the string can be found from what is given.

3. Show that the principle of Virtual Velocities holds good in the case of equilibrium described in the previous question.

4. A weight W, of 34lbs., suspended from a string MD, which is attached to a fixed point M, rests at D on the inclined plane CA, whose base AB is 39 inches; and its vertical height CB, 26 inches. The position of D is such that AD is at right angles to CA; and the point M is in BC produced, CM being 8 inches. Prove that the tension of the string MD is 20lbs.; and find the reaction of the plane on the weight.

5. A uniform heavy rod AB, 14 feet in length, weighing 56lbs., and having a weight W, which is also 56lbs., suspended from a point E in the rod, is sustained in a horizontal position by strings AD and BC, which are attached to fixed points D and C in the same horizontal line DC. The length of AD is 15 feet; of BC, 13 feet; and the perpendicular distance between the parallel lines DC and AB is 12 feet. Show that the point E, from which the weight is suspended, divides the rod in the ratio of 3 : 11.

NORMAL SCHOOL.

FINAL EXAMINATION.

JUNE 1873.

1. Three smooth vertical posts are fixed at the angles of an equilateral triangle and a cord is passed round them, to each end of which a force of 100lbs. is applied; find the pressure on each post.

2. A beam, AB has one end attached to a hinge A, and the other end attached to a cord BC, one end of which is tied to a peg. The weight of the beam is 50lbs.,

and may be supposed to act at its middle point. The beam and cord make angles of 60° on opposite sides of the vertical. Find the tension of the cord.

3. Define the *centre of gravity* of a body, and determine the position of the centre of gravity of a triangular board.

If the centre of gravity of a four-sided figure coincide with one of its angular points, show that the distance of this point and the opposite angular point from the line joining the other two angular points are as 1 to 2.

4. A uniform iron rod, a foot of which weighs $1\frac{1}{2}$lbs., rests on a fulcrum two feet from one end; find what weight, suspended from that end, will keep it at rest when the pressure on the fulcrum is 150lbs.

5. An equilateral triangle is placed upon an inclined plane, its lowest angle being fixed: find how high the plane may be elevated before the triangle rolls.

6. In a system of pulleys, in which each pulley hangs by a separate string, there are three pulleys of equal weights; the weight attached to the lowest is 32lbs., and the power is 11lbs.; find the weight of each pulley.

SECOND CLASS CERTIFICATES.

December 1873.

1. Describe the wheel and axle, and state the mechanical advantage of the machine. Show that the wheel and axle, in the state of equilibrium, is identical with the lever.

2. ACB is a straight lever, whose fulcrum is C. Find the weight which, acting at A, is balanced by $15\frac{1}{2}$ oz. at B, and, acting at B, is balanced by $16\frac{3}{8}\frac{6}{7}$ oz. at A.

3. A straight lever ACB, whose fulcrum is C, is kept at rest in a horizontal position by two forces, one a weight suspended from B, and the other the tension of a string which is attached to A, and runs in the direction AE. If AE be equal to AC, and the perpendicular distance of E from AB be equal to BC, find the relation between the weight suspended from B and the tension of the string EA.

4. Draw a diagram representing a system of pulleys (3 moveable pullies in the system), in which the mechanical advantage (that is, $\frac{W}{P}$) is 8.

(*b*) If $W = 16$lbs, state the tension of the string passing round the middle pulley.

5. Let ABCD be a square. Bisect BC in E and CD in F. Join AE and AF. If a particle A be acted on by two forces, represented in magnitude and direction by the lines AE and AF respectively, find the numerical value of the length of a line representing the magnitude of their resultant, the sides of the square being unity.

FIRST CLASS CERTIFICATES.

DECEMBER 1873.

1. Let ABCD be a parallelogram such that each of the four sides is equal to the diagonal AC. Bisect AC in E. A particle at A being acted on by three forces represented in magnitude by AD, $(1 + 2\sqrt{\frac{1}{3}})$ AE, and $\frac{1}{2}(1 + \sqrt{\frac{1}{3}})$ AB, respectively, and in direction by AD, EA (not AE), and AB, respectively, prove that their resultant is in the direction of the diagonal of a square described on AC.

2. A uniform heavy rod, AC (which may be considered as a line), is in equilibrium, in a horizontal position, with one end C resting on a smooth inclined plane ED, and the other end A connected by a string AB with a fixed point B. Prove that the tension of the string and the reaction of the plane on the rod, are equal to one another.

3. Let BCD be a triangular board, forming two opposite inclined planes, BD being horizontal. Suppose a perfectly flexible uniform heavy cord ABCDE, weighing 14lbs., to pass over it, the portions BA and ED hanging vertically. Let the lengths of AB, BC, CD, and DE, be 6 ft., 8 ft., 6 ft., 8 ft., respectively. The cord being supposed capable of moving freely in virtue of its own weight, inquire what weight (if any) must be attached to either extremity (A or E as the case may be) in order that equilibrium may subsist.

SECOND CLASS CERTIFICATES.

July 1874.

1. ACB is a bent lever, the fulcrum being C. The arms AC and CB are straight; and the angle formed by BC and AC produced is $\frac{2}{3}$ of a right angle. If a weight of 10lbs. acting at A balance a weight of 8lbs. at B when the arm AC is horizontal, find what weight at A will balance a weight of 8lbs. at B, when BC is horizontal.

2. (*a*) Enunciate the principle of Virtual Velocities.

(*b*) Assuming the principle, apply it to determine the relation between the Power and the Weight in the Wheel and Axle.

3. (*a*) Draw a diagram representing a system of pulleys (three movable pulleys in the system) in which a separate string passes round each of the pulleys.

(*b*) Apply the principle of Virtual Velocities to determine the relation between the Power and Weight in this system of pulleys.

4. (*a*) Enunciate the principle of the Parallellogram of Forces.

(*b*) If a particle at A be acted on by two forces represented in magnitude and direction by the lines AB and AC respectively, CAB being a right angle; if the length of AB be 5 feet, and the length of the resultant of the two forces be 13 feet; find the length of AC.

FIRST CLASS.

1874.

1. A uniform heavy rod AB, whose weight is 20lbs., is kept at rest in a horizontal position by four forces (in addition to its own weight): a force of 10lbs., acting at B in a direction BD at right angles to AB; a force of 10lbs. acting at A in a direction AE which is such that EAC is $\frac{1}{3}$ of a right angle; a force m acting vertically at C; and a force n acting horizontally at B. If $AC = 2CB$, find m & n.

2. On AB, an inclined plane, whose base is AC, and which has the angle BAC equal to $\frac{2}{3}$ of a right angle, a heavy body is kept at rest by two equal forces, the one

acting in the direction of AB, and the other in a direction parallel to AC and *towards the plane*. Prove that the reaction of the plane on the body is equal to the weight of the body.

3. Let ABCD, a uniform heavy square, be suspended by a string EB from a point E; and from A and C let weights P and Q be suspended by strings AP and CQ. Prove that if P exceed the weight of the square by 3Q, the direction of the string EB will bisect the line drawn from A to the centre of the square.

SECOND CLASS CERTIFICATES.

July 1875.

1. A straight lever ACB, without weight, the fulcrum being C, is in equilibrium, in a horizontal position, under the influence of two weights, namely, P acting at A, and W at B. If $AC = 3\frac{1}{3}$ feet, and $BC = 4\frac{2}{3}$ feet, and if the pressure on the fulcrum be 24lbs., find P and W.

2. Assuming the Principle of Virtual Velocities, apply to it to determine the mechanical efficiency $\left(\frac{W}{P}\right)$ of the Screw, the Power being supposed to act in a plane at right angles to the direction of the screw.

3. A weight W is kept at rest on an inclined plane by a power A acting parallel to the base. Does such a machine ever act at a mechanical disadvantage? If so, when? Illustrate by an example.

4. If a pupil should say that in the case of a body at rest on an inclined plane under the influence of a Power P acting parallel to the plane, the weight of the body and the force P, being neither equal nor directly opposite, cannot possibly counterbalance one another; and should ask what force, additional to these, acts on the body so as to keep it at rest; what would you reply?

5. What is meant by the *Resultant* of a number of forces acting at a point?

Draw any lines AB, BC, CD; and let a particle at A be acted on by forces parallel to the lines AB, BC, CD, taken in order, and represented by them in magnitude.

Prove (assuming the principle of the Parallelogram of Forces) that the resultant of these three forces is represented in direction and magnitude by AD.

6. ABC is an equilateral triangle, of which the side is one foot. A particle at A is acted on by a force represented in magnitude and direction by AB. Let the force be resolved into two forces, one in a direction parallel to BC, the other in a direction perpendicular to BC. Find the lengths of the lines representing these forces respectively.

FIRST CLASS CERTIFICATES.

JULY 1875,

1. A particle at A, a point in the straight line FAE, is at rest under the influence of three forces, namely, a force of 1lb. acting in the direction AC, which makes the angle FAC equal to one third of a right angle; a force of m lbs, in the direction AB, which is at right angles to AC, and makes BAE two-thirds of a right angle; and a force of n lbs., in the direction AD, making the angle DAE one-half of a right angle. Find the relation between m and n.

2. Let AB be a uniform heavy beam, resting with one end B against a wall, and the other end A on the ground. If the reaction of the wall on the beam, and the friction at B, be together equal to the reaction of the ground on the beam at A, compare the distance of A from the wall with the height of B above the ground.

3. Let ABC be a uniform straight rod, in a horizontal position; AB being $6\frac{2}{3}$ feet; and BC, $3\frac{3}{5}$ feet. In DB, a straight line drawn at right angles to AC in a vertical plane, take the point D above the rod, and let DB be $4\frac{1}{3}$ feet. Suppose the rod to be acted on by two forces besides its own weight, namely, a force of 6lbs. acting at A in the direction AD, and one of 8lbs. acting at C in the direction CD. If the rod weigh 10lbs., enquire whether it be in equilibrium. If it be not in equilibrium, specify any force or forces, which, in conjunction with those acting on it, will produce equilibrium. (*It may be assumed that ACD is a right angle*).

4. Let ABC be a horizontal line, and let BE and BD represent two inclined planes lying towards opposite sides, the angle ABE being two-thirds of a right angle, and the angle CBD being one-third of a right angle. Prove, that, if a uniform heavy rod FG, lying on the planes, the extremity F on BD, and G on BE, be in equilibrium, BF is half the length of the rod.

NORMAL SCHOOL.

FIRST CLASS CERTIFICATES.

1875.

1. ABCD is a square. In CD a point E is taken, and in CB a point F, such that each of the angles EAC and FAC is one third of a right angle. Find the magnitude of four forces acting on a particle at A, namely, a force of 3lbs. in each of the directions AD and AB, and a force of $\sqrt{(6)}$lbs. in each of the directions EA and FA (*not* AE and AF).

2. A uniform straight rod AB, with one end A resting on the ground, has its other end B connected by a string with a fixed point C in the vertical line CA. The rod being in equilibrium under its own weight, the tension (T) of the string, friction (F), and the reaction (R) of the ground on the rod, prove that if ABC be an equilateral triangle, $\frac{R}{3}$ is a mean proportional between $T + F$ and $T - F$.

3. Let AC be a uniform straight rod, in a horizontal position; B, its middle point; D, a point above the rod, in the vertical plane passing through the rod; and DE, the perpendicular let fall from D on AC. If the weight of the rod be represented in magnitude by the straight line DE, prove that the rod will be in equilibrium under the influence of the following forces, namely, its own weight, a force acting at A in the direction AD and represented in magnitude by $\frac{AD}{2}$, a force acting at C in the direction CD and represented in magnitude by $\frac{CD}{2}$, and a force acting in the direction of the rod and represented in magnitude by EB.

NORMAL SCHOOL.
SECOND CLASS CERTIFICATES.
June 1875.

1. Give two familiar examples of a lever working at a mechanical disadvantage, and point out the use of such a machine.

2. If a straight lever be in equilibrium in a horizontal position under the influence of two weights, show that it will be in equilibrium under the influence of the same weights applied at the same points, in every position.

3. In the straight lever ABDE, let AB, BD, DE, be equal distances of one foot; and let the lever (supposed to be without weight) be in equilibrium round the fulcrum C, under the influence of a weight of 3 lbs. applied at A, of 5 lbs., applied at B, of 7 lbs. applied at D, and of 9 lbs. applied at E. Find the position of the fulcrum C.

4. Assuming the principle of the Parallelogram of Forces, deduce the corollary, that, if the directions of three forces acting at a point be parallel to the sides of a triangle taken in order, and their magnitudes be proportional to the sides, they will keep the point at rest.

5. A particle at A is acted on by a force in the direction AB. Suppose this force resolved into two forces, one acting in a direction AC, and the other in a direction AD at right angles to AC, the angles BAC and BAD being each less than a right angle. If the resolved part of the force in the direction AC be $1\frac{2}{3}$ of the whole force acting on the particle, inquire what part of that whole force the resolved part in the direction AD shall be.

6. (*a*) Draw a diagram representing a system of two movable pulleys, the last pulley supporting the weight, and the free portions of the strings being all vertical; and state the mechanical advantage $\left(\dfrac{W}{P}\right)$ when the weights of the pulleys are not taken into account.

(*b*) Let the weights of the pulleys be taken into account, the weight of each being Q. Inquire, then, what must be the relation between P and Q in order that no mechanical advantage may be either gained or lost by the machine.

SECOND CLASS CERTIFICATES.

July 1876.

1. What particulars are required to be known in order to specify a force? What are the conditions of equilibrium of three forces acting at a point? What of three forces acting on a rigid body?

2. Forces of 1lb., 4lbs., and 6lbs., respectively, act on a particle, the force of 4lbs. being inclined at an angle of $60°$ to each of the others, find the magnitude and direction of their resultant.

3. (a) What is meant by the moment of a force with respect to a point? State the principle of moments.

(b) A uniform rod a foot of which weighs 3lbs., rests on a fulcrum two feet from one end, what weight suspended from that end will keep it horizontal when the pressure on the fulcrum is 300lbs.?

4. A circular plate of two feet radius has a circular hole of eight inches radius cut in it. find the distance of the centre of gravity from the centre of the plate, if the centre of the plate and hole are twelve inches apart

5. n cylinders of the same height h, the radii of which are equal to $r_1, r_2, r_3, \ldots r_n$, respectively, stand one upon another with their axes in the same straight line; find the height of their common centre of gravity above the base of the first.

FIRST CLASS CERTIFICATES.

July 1876.

1. Enunciate the Triangle of Forces, and by means of it deduce the Principle of Moments.

Find the resultant of three forces acting in consecutive directions round a triangle, and represented respectively by its sides.

2. A lever without weight is c feet in length, and from its ends a weight is supported by two strings in length a and b feet respectively. Find the ratio of the lengths of the arms, if there be equilibrium when the lever is horizontal.

3. A piece of uniform wire is bent into the form of a triangle; find the position of its centre of gravity.

4. A swing-gate weighing 96lbs. rests on a hinge A, and against a frictionless turning-point B, four feet directly beneath A. Find the strain on the hinge and the pressure on the point, given that the centre of gravity of the gate is 4 ft. 7 in. from AB.

What will be the strain and the pressure if a boy weighing 108 lbs. stands on the gate 6 ft. from AB?

INTERMEDIATE EXAMINATION.

June 1876.

1. State the principle of the Parallelogram of Forces. Explain the meaning of the terms employed in your statements. Shew that if four forces acting on a point be represented by the sides of a rectangle taken in order, they will be in equilibrium.

2. Apply the Triangle of Forces to find the least horizontal force necessary to draw a wheel four feet in diameter and weighing 10 cwt. over an obstacle the height of which is six inches, situated on the horizontal plane on which the wheel rests.

3. Define the moment of a force with respect to a given point.

A uniform beam A B, whose weight is 100 lbs. and length 50 feet, rests with one end (A) on a horizontal plane A C, and the other end against a vertical wall C B. If a string C A, equal in length to C B, prevents the beam from sliding, find the tension of the string.

4. In the system of pulleys in which each pulley hangs by a separate string, a platform is suspended from the lowest block; what force must a man who weighs 140 pounds, standing on the platform, exert to sustain himself when there are three movable pulleys;

5. State the principle of *virtual velocities*, and apply it to find the relation between the power and the weight in the inclined plane.

INTERMEDIATE EXAMINATION.
December 1876.

1. A uniform straight bar, 18 feet long, has weights of 12 and 20 pounds at the two ends. How heavy must the bar be, in order that it may rest on a pivot 10 feet from one of its extremities?

2. A bent lever has arms of equal length, making an angle of 120°. Required the ratio of the weights at the ends of the arms when the lever is in equilibrium with one arm horizontal.

3. (*a*) State the principle of the Parallelogram of Forces.

(*b*) A unit of force is resolved into two forces, F and G acting in directions at right angles to one another. If the magnitude of F be $\frac{1}{2}$, what is the magnitude of G?

4. (*a*) State the principle of Virtual Velocities.

(*b*) Draw a diagram exhibiting the system of pulleys in which the same string passes round all the pulleys and the parts of it between the pulleys are parallel.

(*c*) Apply the principle of Virtual Velocities to determine the ratio between the Power and the Weight in this system.

APPENDIX I.

Duchayla's Proof of the Parallelogram of Force.

In Chapter II. we explained the meaning of the proposition known as the "Parallelogram of Forces," and deduced its truth by means of experiment. We shall now establish the truth of the proposition upon evidence of the same kind as that on which the Theorems of Euclid rest.

The demonstration is divided into three parts :

(1) To find the direction of the resultant when the forces are commensurable.

(2) To find the direction of the resultant when the forces are incommensurable.

(3). To find the magnitude of the resultant.

1. To demonstrate the Parallelogram of Forces, so far as relates to the direction of the resultant, the forces being commensurable.

(1) We assume that if two equal forces act on a particle the direction of the resultant will bisect the angle between the directions of the forces. This seems obvious, for no reason can be adduced for its inclining towards one of the forces which would not equally apply to make it incline towards the other.

(2) Suppose that the proposition is true for two forces p and m, inclined at any angle; and also for two forces p and n, inclined at the same angle; we shall that it will be true for two forces p and $m + n$, inclined at the same angle.

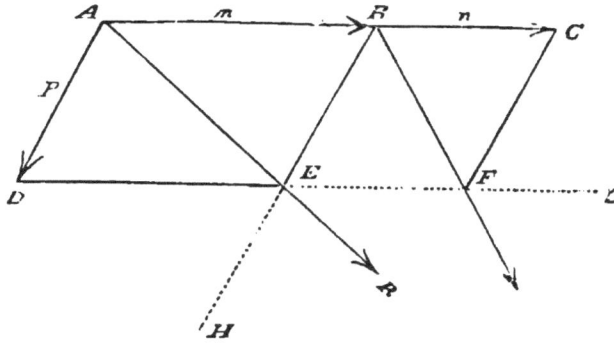

Let A be the particle on which the forces act. Draw AB, AD representing the forces *m* and *p* respectively in magnitude and direction.

Produce A B, and make BC proportional to the force *n* in the same ratio as AB is to the force *m*.

Then since a force may be transferred to any point of its direction, BC will represent the force *n* in magnitude and direction.

Complete the parallelogram ABED, and draw the diagonal A E. Then by hypothesis, A E represents the resultant (R suppose) of *p* and *m* in direction; and the point of application of this resultant may may be supposed to be at E, a point in its direction rigidly connected with A (art. 15).

Now since the force R acting at A is equivalent to two *p* and *m*, when at E it may evidently be resolved back again to the same forces acting parallel to their original directions.

Produce DE to L, and BE to H. Then instead of the forces *p* and *m* at A, we have a force *m* acting along L at E, and a force *p* acting along EH at E.

We may again change the points of application of these forces so that *m* may be supposed to act at F and *p* at B.

Complete the paralellogram EBCF and draw the diagonal B F. Then, by hypothesis, B F represents in direction the resultant of the two forces *p* and *n* acting at P. And this resultant may be supposed to act at F a point in its direction, instead of at A.

At F it may be resolved again into the forces p and n. We have now removed the forces p and $m+n$ which acted at A to F.

Therefore F must be a point in the resultant.

Whence the direction of the resultant of p and $m+n$ is along the diagonal of the parallelogram constructed on the lines representing them.

Now, we know the proposition to be true for the direction of the resultant of two equal forces p and p.

Since therefore it is true for p and p, and p and p, by the preceding it is true for p and $p+p$ or $2p$. And since it is true for p and p, and p and $2p$, it is true for p and $2p+p$ or $3p$, and so on; therefore generally it is true for p and $r p$, where r is a whole number.

And again, since it is true for $r p$ and p, it is true for $r p$ and $2p$, and so on by similar successive deductions it may be proved to be true generally of $r p$ and $s p$, where s is a whole number; therefore it is true for all commensurable forces, i.e. for all forces the ratio of whose magnitudes can be expressed by the ratio of two whole numbers.

Remarks.—The preceding demonstration generally seems obscure to students who meet with it for the first time. This results from the somewhat unusual *form* of the proof. The student will notice that the demonstration consists of two parts. In the first part it is shown that if the principle is true in two cases, viz., with regard to the pair of forces p and m and the pair p and n, it must also hold good in a third case, viz., in regard to the pair of forces p and $m+n$; this part of the proof is purely hypothetical, as much so as in the case of a demonstration by reduction to an absurdity. The second part of the proof takes up the argument, but as a matter of fact the proposition is true in two certain cases; therefore it must be true in a third case, therefore in a fourth case, and so on.

APPENDIX.

(2) *To demonstrate the Parallelogram of Forces so far as relates to the direction of the resultant, the forces being incommensurable.*

Let AB, AC represent in magnitude and direction two of the incommensurable forces.

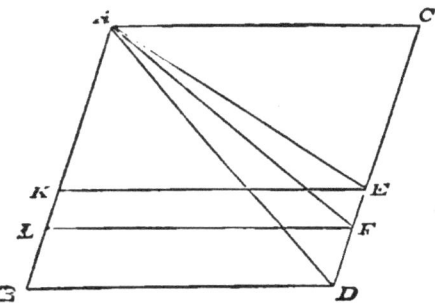

Complete the parallelogram ACDB. Draw AD. Then we have to show that AD represents in *direction* the resultant.

If it do not, let any other line as AE represent that direction.

Divide AC into a number of equal parts, such as that each is less than ED. Mark off along CD parts, each equal to one of these. Then one division will evidently fall between E and D, at F suppose.

Join AF, and draw EK, FL parallel to AC.

The forces represented by AC, AL are commensurable.

Therefore their resultant will be in the direction of AF, and we may suppose this resultant to be substituted for them.

Then the resultant of the forces represented by AC and AB is equivalent to the resultant of two forces, one acting in the direction AF, the other represented by LB, and which may therefore be supposed to act at A, in the direction AB; and this resultant must lie *within* the angle BAF.

But, by hypothesis, it acts in the direction AE, *without* the same angle, which is absurd.

In like manner it may be shown that no direction but AD can be that of the resultant of the forces represented by AB, AC.

Whence the proposition is true for *all* forces as far as the *direction* of resultant is concerned.

(3) *To demonstrate that the Parallelogram of Forces holds also with respect to the magnitude of the resultant.*

Let A C, A B represent in magnitude and direction two forces P and Q.

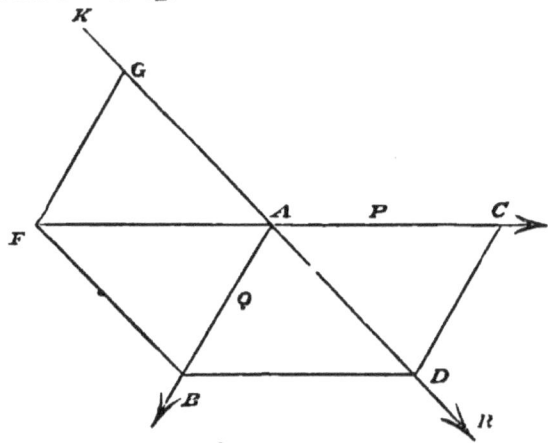

Complete the Parallelogram B C.

Then by what has already been proved, the diagonal A D will be the direction of the resultant of P and Q.

Produce A D backwards towards K.

Then if we suppose a force equal in magnitude to the resultant of P and Q to act along A K, this force and P and Q will be in equilibrium But if any number of forces be in equilibrium any one is equal and opposite to the resultant of all the rest. Thefore the force P must be equal and opposite to the resultant of the force acting along A K and Q.

Produce C A to F, and make A F equal to A C.

Then AF represents in magnitude and direction **the** resultant of Q and the force along A K.

Join F B, and draw F G parallel to A B.

Then since FAC is parallel to BD, and FA=AC=BD, therefore also FB is equal and parallel to AD (Euc. I. 33).

Now AF represents in magnitude and direction the resultant of Q and the force acting along DK. But the forces represented in magnitude and direction by the lines AG, AB, have a resultant acting along AF. Therefore AG must represent in magnitude the force which we supposed to act along AK. But this force is equal and opposite to the resultant of P and Q

Therefore the magnitude of the resultant of P and Q $= AG = FB = AD$.

That is, the magnitude of the resultant of P and Q is represented by the diagonal of the parallelogram, of which the sides represent the magnitude of the Forces P and Q

Therefore, if two forces, &c., Q E. D.

RESULTS, HINTS, &c., FOR THE EXERCISES.

Chapter I.

Exercise I. Page 6.

(4) A force is measured numerically by the weight it would sustain if acting vertically upwards. The most common unit of weight employed is 1lb. Hence in the absence of other information, I should understand by a force of 10 or a force of P, a force of such magnitude that it would support 10lbs. or Plbs. respectively if applied to the weight in the manner stated. (8) 8lbs.

PARALLELOGRAM OF FORCES.

Chapter II.

Exercise I. Page 12.

(2) 15lbs. in the same direction as the forces.
(3) 1lb. in the direction of the greater force.
(4) 11 oz. acting in the direction of the latter forces
(8) 15lbs. (9) 13lbs. (10), If P be smaller force $R = P\sqrt{10}$.
(11) 6lbs. (12) 30lbs. (13) 1clbs. (14) the latter.
(15) 27lbs. and 120lbs. (16) 1.2lbs. and 1.6lbs.
(17) 129.7lbs. and 129.5lbs.

Exercise II. Page 16.

(1) 61 lbs. (2) 7 lbs. (3) 7 lbs. (4) 305 lbs.
(5) Let the force along each rafter be P, then we have two equal forces, each P, making an angle of 60° with each other, and having for resultant 90 lbs.; hence $P = 30\sqrt{3}$. (6) 4.63 lbs. (7) 31.90 lbs.
(9) This is the same as to find the resultant of two equal forces, each equal to 6 lbs., making an angle of 60° with each other. The resultant is $6\sqrt{3}$ lbs. (10) First, we have given the resultant, W, of two equal forces, acting at an angle of 60°, to find the forces. If P be one of the equal forces, $W^2 = (P + \frac{P}{2})^2 + (\frac{P}{2}\sqrt{3})^2$, or $P = \frac{W}{\sqrt{3}}$. The pressure on B or C is resultant of two equal forces, each $= \frac{W}{\sqrt{3}}$, acting at an angle of 120°; resultant is, therefore, equal to either of the forces, that is pressure on B or $C = \frac{W}{\sqrt{3}}$. Pressure on A is resultant of two equal forces, $\frac{W}{\sqrt{3}}$, acting at angle of 60°; pressure on A is, therefore, W.

Exercise III. Page 17.

(3) The resultant is equal to one of the forces. (4) 120°.
(5) Euc. I 20, (6). If P be the lesser force, resultant will be 2P, and greater will be $P\sqrt{3}$; therefore sides of triangle are in ratio of 1, 2, $\sqrt{3}$; hence angle between resultant and lesser force is 60°, and between resultant and greater 30°. Or thus, bisect resultant and join point of bisection with right angle. In any right-angled triangle the line joing the point of bisection of the hypothenuse and the right angle is half the hypothenuse. The angle between lesser fc. and resultant, is, ∴ an angle of an equil. triangle.

(7) A force AD. (8) $\sqrt{2}$ at C, parallel to DB.
(10) $6\sqrt{2}$. (12) The straight lines drawn from the angles of a triangle to the points of bisection of the opposite sides meet at the same point and each of the bisecting lines is trisected at that point (art. 78).

TRIANGLE OF FORCES.

CHAPTER II.

Exercise I. Page 21.

(1) Euc. I, 20. (2) Arts. 29 and 31. (5) Euc. I, 48.
(6) The side which does not pass through the point, represents the third force in direction, but not in line of action. (7) Construct triangle having its sides 5, 6, 7. Produce side 7 backwards through intersection of 5, 7, and make produced part equal to side produced. Through intersection of 5, 7, draw a line parallel and and equal to 6. The lines 5, 6, 7, acting at angular point will represent forces in direction and magnitude.
(8) The forces make angles of 60° with each other.
(9) $2P$ making an angle 60° with P. (10) Makes an angle of 30° with greater force, see Ex. 6, page 18.
(11) $1 : \sqrt{2}$.

Exercise II. Page 24.

(1) 8olbs. and 6olbs. (2) 8lbs. and 6lbs. (3) Produce side 24 backwards and const. right-angled triangle having side making angle 30° with vertical for hypothenuse. The three sides of this triangle are parallel to the 3 fcs., which keep point at which wt. is attached at rest, and, therefore, proportional to them. Let hypothenuse $=2$, &c., and horizontal tension is $8\sqrt{3}$, the other $16\sqrt{3}$.
(4) The sides of the triangle whose sides are 10, 6, 8, are parallel to the three forces which keep point B at rest, and, therefore, proportional to them. Tension is $26\frac{2}{3}$ cwt.

(5) Let vertical through C meet line AB in D. Then AB will be bisected in D. Through D draw DE parallel to BC; AC is bisected in the point E. The straight line drawn through point of bisection of one side of a triangle parallel to the base bisects the other side. (This may each be proved by drawing through point of bisection of one side of triangle lines parallel to other two sides. Show that the two triangles thus formed are equal (Euc. 1. 26). Then since opposite sides of parallelograms are equal, we have the required result). The sides of the triangle CDE are parallel to the three forces which keep point C at rest, and therefore proportional to them. $CD = 12$, $AD = 35$, $\therefore AC = 37$; $EC = \frac{1}{2}AC = $ &c. The weights are equal to tension of string $= 154\frac{1}{6}$lbs.

(6) $17 : 8$. (7) 3lbs.

Exercise III. Page 27.

(1) The sphere is kept at rest by reactions of planes and its own wt., which may be supposed to act at its centre. These three fcs. must, therefore, pass through centre (art 34). Show that the angles which line of action of wt. makes with reactions at centre are $60°$ and $30°$. Triangle formed by one of the reactions, weight of sphere, and plane, will have its side parallel to three forces which keep sphere at rest &c. Pressures are $100\sqrt{3}$ and 100lbs. respectively.

(2) Through top of h draw line parallel to F. Triangle formed will have its side parallel to the three forces F, W, and reaction of h which passes through centre.

(3) Produce the wt. backwards till it meets the tension; then reaction of hinge must pass through this point, tension $= 5\sqrt{2}$lbs. (4) 50lbs. (5) Produce wt. backwards till it meets reaction of wall; join this point with A. The triangle of which this line is hypothenuse has its sides parallel to the three forces which keep rod at rest. If perpendicular of triangle be 1, base is $\frac{1}{2}$, and, therefore, hypothenuse $\frac{1}{2}\sqrt{5}$. Hence $W : R :: 1 : \frac{1}{2}\sqrt{5} \therefore R = \frac{W}{2}\sqrt{5}$.

(6) Beam is kept at rest by three forces, W acting at its middle point, T, tension of string, and reaction of wall. These three meet in string CB, at a point E. BC is bisected in E, (note Ex. 5, Exercise II). From E draw BD perpendicular to wall; then CA = AD, since EA is drawn through E parallel to BD. Let BD = y, CA = x, then AD will also = x. $5^2 = y^2 + (2x)^2$; $3^2 = y^2 + x^2$. Therefore $x = \sqrt{\tfrac{4}{3}}$. The sides of triangle CBD are parallel to three forces which keep beam at rest, ∴ T : W :: 5 : 8√3 or $T = \dfrac{5\sqrt{3}W}{8}$.

RESOLUTION OF FORCES.

Chapter IV.

Exercise I. Page 34.

(1) 1.7 and 1.35√3. (2) Draw one side at right angles to 12, and equal to 25; complete parallelogram having 25 for one side and 12 for diagonal; the other side will be 27.73. (7) Resolve tension of cord into two parts, one in direction of keel, and other perpendicular to it. The latter tends to turn the bow towards the land, and this tendency will be counteracted by the pressure on the rudder, if the rudder be turned from the land.

(8) 50√3 lbs. and 50 lb. (9) If P = given force, the first force = P, the second P√2. (10) 40√3 lbs. and 40 lbs. (11) 52 lbs. (12) 25 lbs. (13) Resolve in direction of force 100 and at right angles to it. R makes an angle of 90° with 100, and is = √3.

(14) Resolving the 80 and 100 horizontally we have 40 and 50√3, respectively. The angle between these forces is 30°, and their resultant is found by the parallelogram of forces to be 122.8 lbs.

Exercise II. Page 39.

(1) 2 lb. (2) $12\sqrt{3}$ lbs. (3) $5\sqrt{2}$ lbs. (4) $\sqrt{3} : 1$.
(5) $8\sqrt{3}$ lbs. and $16\sqrt{3}$ lbs. (6) $1 : \sqrt{3}$. (7) Resolve the 20 and $20\sqrt{3}$ along and at right angles to the direction of the force 40, and we have $R = 80$ acting in a direction opposite to the force of 40 lbs.

(8) Resolving along the plane and at right angles to it we have

$$\frac{\sqrt{3}}{2} W = P + \tfrac{1}{2}P, \text{ or } P = \frac{\sqrt{3}W}{3}.$$

$$R = \tfrac{1}{2}W + \tfrac{1}{2}P\sqrt{3}$$
$$= \tfrac{1}{2}W = \tfrac{1}{2}\frac{\sqrt{3}}{3}\sqrt{3}$$
$$= W.$$

(9) Resolve along plane, $\tfrac{1}{2}W\sqrt{3} = 2P$ or $P = \tfrac{1}{4}W\sqrt{3}$.
   „   R, and $R + \tfrac{1}{2}P\sqrt{3} = \tfrac{1}{2}P\sqrt{3} + \tfrac{1}{2}W$
   or $P = \tfrac{1}{2}W$.

(10) Let $t =$ tension in AC, $t' =$ tension in BC.
Resolve horizontally, and $\dfrac{t}{\sqrt{2}} = t'$.
   „   vertically „ $\dfrac{t}{\sqrt{2}} = W$;
$\therefore t = W\sqrt{2}$, and $t' = W$.

(11) The tension of the string must be the same throughout. Resolve 20 and W along each plane, and equate results, and we have $W = 10\sqrt{6} = 24\tfrac{1}{2}$ lbs. nearly.

PARALLEL FORCES.

Chapter V.

Exercise I. Page 46.

(1) $70\tfrac{7}{12}$ lbs. and $50\tfrac{5}{12}$ lbs. (2) 38 lbs. and 114 lbs.
(3) $6\tfrac{7}{18}$ ft. from one end. (4) 12 in.
(5) Resolve 12 lbs. into forces of 4 lbs. and 8 lbs. acting at the extremity of the line bisecting its width; resolve then into 4, 4, ; 2, 2, acting at its corners. (6) 9 lbs. at A, 6 lbs. at B, 4 lbs. at C, 5 lbs. at D.

(7) Resolve the 150lbs. into two parallel forces, one of 50lbs. acting at angular point, and another of 100lbs. acting at middle point of side of triangle. The 100lbs. can be resolved into two equal parallel forces of 50lbs. each acting at angular points. Hence, pressure on each prop is 50lbs.

(8) Pressure on A = 14lbs., B = 6lbs., C = 10lbs., D = 30lbs.

(9) No. The weight may be resolved into 4 equal parts acting on each leg of the table.

(10) Art. 49.

(11) If $x =$ the distance from 2lbs., then, by article 46 we have
$$2x = 5(x - 1\tfrac{1}{2}),$$
$$\text{or } x = 2\tfrac{1}{2};$$
that is, the hand must be placed 1ft. from the middle of the sash or 6in. from the broken cord. Or, we might proceed as follows:—Resolve the 5lbs. into two parallel forces of $2\tfrac{1}{2}$lbs. each, acting vertically downwards, 2lbs. of one of the forces is balanced by the tension of the unbroken cord which is 2lbs. The problem then is, to find the position of the resultant of forces, $\tfrac{1}{2}$lb. and $2\tfrac{1}{2}$lbs 3 feet apart, which gives the same result as above.

MOMENTS.

CHAPTER VII.

Exercise I. Page 56.

(1) Arts. 51, 52. (2) Art. 53. (3) Art. 55, second case, and art. 54. (4) Art. 58; take moments about pt. of intersection of two of the forces, then moment of these two will be *zero*; moment of third must also be zero if the forces are in equilibrium, and must, therefore, pass through same point. (5) Let W be wt. of bundle, b its distance from his shoulder; P, the pressure of hand, a its distance from shoulder; then,
$$Wb = Pa.$$
The left-hand side being constant, if a diminishes P increases, and therefore P + W, or the pressure increases.

Regarding the man and things carried as one body, we see that the change only affects the mutual actions at the head and shoulder, and not the pressure on the ground.

(7) Art. 57. (8) When three forces are in equilibrium each is equal and opposite to the resultant of the other two, then Art. 56. (9) Arts. 59 and 61.

Exercise II. Page 60.

(1) 7 lbs. (2) If the pressure on one prop is 8 lbs., the pressure on the other will be 48 lbs. Let $x =$ distance at which 56 lbs. must be placed; take moments about that point, and we have

$$48 x = 8 (7 - x);$$
$$\therefore x = 1 \text{ ft. from one end.}$$

(3) 20 inches from one end. (4) Distance from first man must equal $\frac{1}{3}$ of pole. (5) 1 inch from fulcrum on side of 7 lbs. (6) Take moments about fulcrum, and distance $= 1.74$ ft. (7) 8 inches from smaller force.

(8) Each man pushes *back* the boat with his feet with a force exactly equal to that which he applies to the handle of the oar, *i.e.*, 50 lbs.; but he presses it *forward* also with the force which his oar produces on the rowlock; if P be that force, then $P \times 7\frac{1}{2} = 50 \times 10$, or $P = 66\frac{2}{3}$ lbs. Each man, therefore, presses forward the boat with a pressure equal to $66\frac{2}{3} - 50 = 16\frac{2}{3}$ lbs. The resultant force propelling the boat is $16\frac{2}{3}$ lbs. $\times 8 = 133\frac{1}{3}$ lbs.

(9) 7 lbs. (10) $10\sqrt{3}$ lbs. (11) $P : Q :: 1 : 2$. (12) $1 : 2$.

(13) $P = 30$ lbs.; pressure on the fulcrum $= 36.61$; produce horizontal and vertical forces till they meet, and join their point of intersection with the fulcrum; the pressure on the fulcrum will act along this line (art. 34).

Exercise III. Page 65.

(1) 32 lbs. (2) 1. Reaction of wall $= 83.23$ lbs.; 2 Tension of string. $= 93.23$ lbs.; 3. Reaction of plane $= 112$ lbs.

(4) Produce reactions of wall and ground till they meet. Take moment about this point, and force will $= 10\sqrt{3}$ lbs.

(5) Reaction of wall is equal to horizontal force = W. Take moments about foot of ladder, and required distance = $4\frac{1}{2}$ ft.

(6) The weight of each beam, acting at its centre, may be resolved into 50lbs. acting at A, and 50lbs. acting at C. Take moments about C, and thrust = 50lbs.

(7) Since the ladder lies on the wall, the reaction of the wall will be at right angles to the ladder. (Art. 62, III). Produce reactions of wall and horizontal plane till they meet. Take moments about points of intersection; then if P be the pressure on the peg, $P \times 4\sqrt{2} = 120 \times \frac{5}{3} \times \frac{1}{\sqrt{2}}$, or $P = 25$lbs. To find reaction of wall, take moments about foot of ladder, and reaction = $25\sqrt{2}$.

Exercise IV. Page 67.

(1) From D draw perpendicular p, p', on each of equal sides a; then $a \times p + a \times p' = 2$ Area triangle ABC $= a \times P$.

∴ $p + p' = P$. If F be one of the equal forces.
$p \times F + p' \times F = P \times F =$ const.

(2) Sum of perpendiculars from any point within an equilateral triangle = perpendicular from angular point on opposite side.

(4) The line drawn from the middle point of the hypotenuse to the right angle is half the hypotenuse.

(5) Since the force acting along the rope is constant, its moment or turning power will be greatest when the perpendicular on it from the foot of the tree is greatest. The perpendicular will be greatest when the point of the tree to which the rope is attached is equal to the distance of the man from the foot of the tree. This can easily be shown by a simple geometrical construction. It is equivalent to saying that if a rectangle and a square have equal diagonals, the perpendicular from an angular pt. of the square on its diagonal is greater than corresponding perpendicular in rectangle. Or, we may reason as follows:

RESULTS, HINTS, &C.   179

Let a = length of each side of right-angled triangle when sides are equal; p = corresponding perpendicular; l = length of rope.

Let b, c, be sides when unequal, and p' = corresponding perpendicular; l will be the same in each case.

Then $aa = pl$; $bc = p'l$, or $\dfrac{a^2}{bc} = \dfrac{p}{p'}$; also

$b^2 + c^2 = l^2$, and $2a^2 = l^2$; $\therefore 2a^2 = b^2 + c^2$.

Hence $\dfrac{b^2 + c^2}{2bc} = \dfrac{p}{p'}$; but the sum of the squares of two quantities is greater than twice their product. Therefore p is greater than p'.

CENTRE OF GRAVITY.

CHAPTER VII.

Exercise I. Page 72.

(1) 12 inches from 20lbs. (2) Between the 2lbs. and 3lbs., and 8 inches from the latter. (3) 10 inches from the 12lbs. (4) 6 inches. (5) 2 inches. (6) 6lbs. (7) 9 feet. (8) Let W be wt. of rod acting at its C. G. Take moments about edge of table and W × 1 = 1 × 2, or W = 2lbs. (9) 8½ inches from the 7lbs. (10) The 60lbs. acts at C. G. Each man bears 30lbs. Let x = distance of other man from C. G. Take moments about C. G., and $30x = 30 \times 8$, or $x = $ 8ft. from C. G. or 4ft. from other end. (11) 1ft. from smaller end. (12) W × 1 = 3 × 4, or W = 12 cwt. (12) If x = distance of C G. from top, then
$6x = 4 \times 3.39 + 2 \times 3$, or $x = 3.26$; that is the C. G. is 3.26 inches from the top.

(13) Articles 50 and 65. (14) Art. 65.

Exercise II. Page 76.

(1) Articles 69 and 70. (2) Art. 75. (3) Art. 73. (4) Art. 72. (5) Arts. 73 and 74. (6) The vertical line through C. G. falls within the base. (7) When at greatest height vertical through C. G. falls at end of

diameter of base. If slant height be 5 and height 4, diameter of base must be 3: but diameter of base is 60; therefore the height is 80ft. (8) $2\sqrt{3}$. (9) Art. 76. (10) The C. G. is raised, and returns to its original position. (11) When it rests on two pts. if it receive a small displacement, the C. G. will descend and cannot return to its original position. (12) To upset a body in a state of stable equilibrium we must raise its C G. The solid cylinder will be the heavier, *cæteris paribus*, and will therefore require a greater force to upset it. (13) If there were only two feet, the equilibrium would be unstable. (14) Because the base is so small that it is practically impossible to place the pin so that the line of direction may fall within it. But even if it were so placed it would not stand for an instant, for its altitude is so great in proportion to the base, that the slightest force would be sufficient to upset it.

(15) The C. G. is higher in the first case, and therefore the equilibrium is more unstable.

(17) 4 inches. (18) It is in stable equilibrium when its C. G. is directly under its geometrical centre. It is in a position of unstable equilibrium when its C. G. is vertically above its geometrical centre. (20) 60°. (21) The C. G. is in the plumb line which, therefore, bisects the hypothenuse, and is also equal half of it. The acute angles are 60° and 30°. The ratio of the sides are as $\sqrt{3} : 1$.

Exercise III. Page 85.

(1) Let wt. of 2lbs. be suspended at C, in the triangle ABC. Bisect AB in D; join DC, and bisect it in E. E is the C. G.

(3) Let a = length of either of the equal sides. The C. G. of squares will be at right angle. Take moments about this point

and $$\left(2a^2 + \frac{a^2}{2}\right)x = \frac{a^2}{2} \times \frac{2}{3} \cdot \frac{a}{\sqrt{2}};$$

$$\therefore x = \frac{a\sqrt{2}}{15},$$ where x is distance of C. G. from vertex.

(5) 15in. from side corresponding to wts. 3, 4; 12in. from side corresponding to wts. 4, 6. From centre, therefore, $\sqrt{5}$ inches.

(6) $\frac{1}{6}a$ from centre. (7) On diameter passing through both centres; to left of centre of large circle at a distance $= \dfrac{r^2 d}{R^2 - r^2}$.

(8) Draw diameter passing through vertex of triangle and bisecting opposite side. Greatest wt. will be at extremity of this diameter. Take moments about point of bisection of side by diameter, and wt. will be found to be equal to wt. of table.

(9) 405.6lbs. (10) Let W be wt. acting at C. G. of slab. Resolve W into $\frac{1}{3}$W at vertex, and $\frac{2}{3}$W at pt. of bisection of the base. Again, resolve $\frac{2}{3}$W into $\frac{1}{3}$W, & $\frac{1}{3}$W, at extremities of base.

Exercise III.   Page 86.

(12) Let a, b, be the sides, then the hypothenuse will be $\sqrt{(a^2 + b^2)}$. Weights at angular points will be ma^2, mb^2, $m(a^2 + b^2)$. Take moments about right angle, and if x be distance of C. G. from right angle

$$2(a^2 + b^2)x = (a^2 + b^2)\frac{ab}{\sqrt{(a^2 + b^2)}};$$
$$\therefore x = \frac{ab}{2\sqrt{(a^2 + b^2)}}.$$

Or, find C. G. of weights ma^2, mb^2; then C. G. of whole will be at middle point of line joining this point with right angle.

(13) The weights of the beams act at their middle points, and are proportional to their lengths. Since hypothenuse is 20, each of equal sides will be $\dfrac{20}{\sqrt{2}}$ or $10\sqrt{2}$. The resultant of these two is $20\sqrt{2}$ acting at middle point of perpendicular. Take moments about right angle, and $(20 + 20\sqrt{2})\ x = 20 \times 10 + 20\sqrt{2} \times 5$, or $x = 5\sqrt{2}$.

(14) Let a = side of square. Take moments about centre of square, and $(a^2 + \frac{1}{2}a^2) x = \frac{1}{2}a^2 \times \frac{3}{8} a \therefore x = \frac{5}{18} a$, the distance of centre of square measured on line joining this centre with vertex of triangle.

(16) 10 cent pieces. (17) Centre of sphere of which hemispherical end forms a part

(18) Art. 79; $\frac{1}{3} (h_1 - h_2)$ from base. (20) Art. 72.

MECHANICAL POWERS.

Chapter VIII.

BALANCES. Sec. II.

Exercise I. Page 94.

(5) The tradesman loses $\frac{1}{132}$lb. or $\frac{55}{132}$ cents. (6) 8lbs. (7) True weight = 6lbs. Let a be shorter arm, b longer arm; then $a+b = 3\frac{3}{4}$ ft.; also $6a = 4b$. Substitute these value and $b = $ 2ft. 9in, $a = $ 1ft. 6in. (8) 90 cents. (9) 20 per cent.

(10) When w is put into the scale of shorter arm, let p be the weight of the goods put into scale-pan of longer arm; then $5p = 4w$, or $p = \dfrac{4w}{5}$. Also, let q from shorter arm balance w from longer, then $4q = 5w$, or $q = \dfrac{5w}{4}$.

In latter case seller's loss will be $q - w = \dfrac{w}{4}$.

In former case seller's gain will be $w - p = \dfrac{w}{5}$.

Therefore seller's loss is $\dfrac{w}{4} - \dfrac{w}{5} = \dfrac{w}{20}$ in two sales, or $\dfrac{50w}{20}$ in 100 sales. Therefore seller's loss $= \dfrac{50}{20} = 2\frac{1}{2}$ per cent. Or since he loses $\dfrac{w}{20}$ on two sales, he will lose $\dfrac{w}{40}$ on one sale, that is $2\frac{1}{2}$ per cent. as before.

(11) $\frac{25}{28}$

THE STEELYARD.

Exercise II. Page 97.

(4) 7 inches from point of suspension. (6) The point from which the graduation begins is brought nearer to the point of suspension C. (7) The point D is taken farther from C.

THE WHEEL AND AXLE,

Exercise I. Page 99.

(2) Art. 99. (4) If t be half thickness of rope, then P : W :: $r+t$: R+t. (5) The rope coils on itself and increases the radius of the axle. (6) 400lbs. (7) 1ft. (8) 3ft. 4in.

THE PULLEY.

Exercise I. Page 105.

(2) 1℔. 12 oz. (3) 4 pulleys. (4) 4 pulleys.
(5) 4 pulleys. (6) $18\frac{1}{4}$ oz. (7) 72 oz.
(8) 8lbs. (9) The weight supported is 160lbs., and the power exerted is 5lbs. (10) 140lbs.

(11) The number of pulleys has been omitted in the question. The system consists of three movable pulleys. Then with three pulleys a power of 1lb. would sustain a weight of 8lbs. 180lbs. must, therefore, be divided into two parts in the ratio of 1 : 8. The power will be 20lbs.

Exercise II Page 107.

(1) w must be less than 5P. (2) 28lbs.
(3) 120lbs. (4) 840lbs. (5) 5 pulleys. (6) 16 strings.
(7) 156lbs. (8) 30lbs. (9) 37lbs.

(10) Since $n\text{P} = \text{W}+w$, $\therefore n - \dfrac{w}{\text{P}} = \dfrac{\text{W}}{\text{P}}; \dfrac{\text{W}}{\text{P}} = 0$ when $n - \dfrac{w}{\text{P}} = 0$ or $n\text{P} = w$. That is, when the weight of the lower block is equal to the power multiplied by the number of strings at the lower block.

(11) Since there are six strings at the lower block, the weight must be six times the power; we have, therefore to divide the 140lbs. into 7 parts, 6 of which will be weight and 1 power. The tension will be 20lbs.

(12) $\frac{1}{3}$ of his weight. (12) A force equal to his own weight.

Exercise III. Page 111.

(1) The number of strings must be two. (2) 8lbs. (3) 440lbs. (4) 841.5lb. 5) Let w be the weight of each pulley. Proceed as in example 5, and put $P = 0$; then $w : W :: 1 : 247$. (6) In the Third System the weights of the pulleys act *with* the power, in the other two systems *against* it.

THE INCLINED PLANE.

Exercise I. Page 114.

(1) 25lbs. (2) 1120lbs. (3) $12\frac{1}{2}$lbs. (4) $6\frac{2}{3}$lbs. (5) Let x be the part hanging over the top of the plane, then $104 - x$ will be the part resting on it. Hence $\frac{x}{104-x} = \frac{3}{5}$; $x = 39$, and the other is 65.

(6) $6\frac{1}{4}$ inches. (7) $10\frac{1}{2}$ feet. (8) Let x be the height and y the length, then $\frac{x}{y} = \frac{40}{56}$, or $7x = 5y$. Also $y^2 - x^2 = 342$, or $y^2 - \frac{25 y^2}{49} = 342$; hence $y = 485.79$, and $x = 346.99$. (9) 2lbs. (10) $28\sqrt{2}$, and $28\sqrt{2}$.

(11 Let h be the height of the planes, and l, l' their lengths; let P be the tension of the string that supports the weights.

Then $\frac{P}{10} = \frac{h}{l}$, and $\frac{P}{7} = \frac{h}{l'}$; dividing, $\frac{l}{l'} = \frac{7}{10}$.

(12) 32 oz.

Exercise II. Page 117.

(1) $20\sqrt{3}$. (2) $15\sqrt{41}$. (3) 56lbs. (4) $12\sqrt{3}$.
(5) 4. (6) 30°. (7) 8.65lbs. (9) 60°.

THE SCREW.

Exercise I. Page 124.

(1) Article 135. From equation expressing the relation between the power and the weight we see that if the arm be increased the efficiency is increased.
(2) 1 : 62.83. (3) 1716lbs. (4) 1.8lbs.
(5) 25.132. (6) 3.1416. (7) .4398 inches.
(8) $\dfrac{2\pi l}{n}$.

VIRTUAL VELOCITIES.

CHAPTER IX.

Exercise I. Page 131.

(4) 15 lbs. (5) $\tfrac{1}{2}$ inch, (6) 6 inches. (7) 3lbs.
(8) 2ft. (9) 180ft. (10) 7lbs.
(11) If P be one of the equal forces, then
$P \times 137 + P \times 105 = 253 \times 88$ ∴ $P = 92$lbs.
(12) 2 : 3; height 27 inches. (13) 144lbs.

FRICTION.

CHAPTER X.

Exercise I. Page 136.

(1) Arts. 134, 138. (2) 137. If a point be taken in either of the sides containing the angle and a perpendicular be let fall on the other side, the perpendicular of the right-angled triangle thus formed divided by the base is the co-efficient of friction.

(4) Friction acting up the hill, the weight of the body and the reaction of the hill, $\mu = 1$. (6) If the co-efficient of friction be $\dfrac{1}{\sqrt{3}}$, the perpendicular is 1 and base $\sqrt{3}$, and, therefore, hypothenuse 2, hence inclination = 308.

(8) In this case the body is on the point of moving down the plane, and, therefore, friction acts with the power (art. 135). Hence $\dfrac{20 + \frac{1}{4}R}{W} = \frac{5}{13}$, and $R = W \frac{12}{13}$;

Hence $W = 130$lbs. (9) The friction acts with the power. Solution same as last question; $P = 6.4$ oz.

(10) See Ex. 2. 5lbs. (11) From the first part of the question we have $F = W$, that is $\mu W = W$, or $\mu = 1$; but $\mu =$ perp. divided by base. The perp. is, therefore, equal to base; hence angle $= 45°$.

Exercise II. Page 138.

(1) Take moments about the intersection of the two reactions. $\mu = \frac{1}{2}$.

(2) Let $R =$ reaction of pavement, and let W be weight of man and ladder both acting at same point. Take moments about intersection of two reactions, and $\mu R \times \sqrt{3} = W \times \frac{1}{2}$; but $W = R \therefore \mu = \frac{1}{6}\sqrt{3}$. (3) See example 3. $7\frac{2}{9}$ feet.

(4) The ladder is kept at rest by R', reaction of vertical plane, R, reaction of horizontal plane, and friction $\frac{1}{6}R'$, and $\frac{1}{4}R$ respectively. Then $R' = \dfrac{R}{4}$; $R + \dfrac{R'}{6} = W$. Take moments about foot of ladder, and $R'\sqrt{(21^2 - x^2)} + \dfrac{R'}{6}x = \dfrac{Wx}{2} = \dfrac{25}{6}R' \times \dfrac{x}{2}$.

$\therefore x = 9.71$ ft. from foot of ladder.

(5) Let R, R' be reactions of plane and wall, respectively; $\frac{5}{8}R$, $\mu R'$, friction of plane and wall respectively. Resolve W into $\dfrac{W}{2}$, $\dfrac{W}{2}$ acting at top and foot of ladder.

Then we have $R' = \frac{5}{8} R$, $R + \mu R' = \dfrac{3W}{2}$.

Take moments about foot of ladder, and if distance of ladder from wall be a, we have
$$\mu R'a + R'a = Wa = (\tfrac{13}{8} R' + \tfrac{2}{3}\mu R)a;$$
$$\therefore \mu = \tfrac{1}{4}.$$

(6) Same notation as in last. $R' = \dfrac{R}{\sqrt{3}}$, $R + \dfrac{R'}{\sqrt{3}} = 3 + 4$.

Take moments about foot of ladder, and we have

$$15R' + \dfrac{R'}{\sqrt{3}} \times 15\sqrt{3} = 3 \times 5\sqrt{3} + \dfrac{x}{2}\sqrt{3} \times 4;$$

$\therefore \mu = 18.75$, that is, the weight cannot be suspended on a round higher than 18.75 feet from foot of ladder.

APPLICATION OF SIMILAR TRIANGLES.

Chapter XI.

Exercise I. Page 144.

(1) Draw FL, FM parallel to PA and WA respectively. Since the lever is kept in equilibrium by the forces P, W, and the reaction at C, the resultant of P and W must pass through C; hence AC is its direction, and by the parallellogram of forces, P and W must be proportional to AM and AL, respectively, or P : W :: FL : FM. Now, by construction PMF and FLW are both isosceles triangles similar to each other,

therefore   FL : FM :: FW : FP,

or   P : W :: FW : FP.

(2) Let E be end of rod from which W is hung, and T tension of string CB. The system is in equilibrium under action of forces, W acting vertically downwards at E, T acting at B in direction BC, and reaction (R) of hinge at A. Since last must be equal and opposite resultant of T and W acting at D where they meet, AD must be its direction. Hence triangle ADC has its sides parallel and therefore proportional to the three forces, which acting at D keep the whole system in equilibrium, therefore T : W :: CD . AC; also from similar triangles CBA, EBD,

CD : AE :: CB : AB,

$$\therefore T = W \frac{CD}{AC} = \frac{W}{AC} \frac{AE.CB}{AB},$$
$$= W \frac{AE.CB}{AC} \cdot \frac{1}{AB};$$

that is, T varies as $\frac{1}{AB}$, because W, AE, BC, and AC are all constant.

(4) Take moments about intersection of reactions, and $t \times 3 = 2\frac{2}{5}W$; $\therefore t = \frac{4}{5}W$.

(5) Since the heights are proportional to the diameters of the bases, the height of the cavity is to 12 as 6 : 8; hence it is equal to 9 inches. The volume of a cone the height of which is h, and radius r is $\pi r^2 \frac{h}{3}$. Hence volume of solid contained by outer conical surface is $\pi.64$, and that contained by the inner surface is $\pi.27$. Therefore volume of shell is $\pi.37$. The C. G. of whole cone is 3 in. from base; that of cavity $2\frac{1}{4}$ in. from base. Hence

$$\pi.64 \times 3 = \pi.27 \times 2\frac{1}{4} + \pi.37 \times x;$$
$$\therefore x = 3\frac{81}{148} \text{ inches, the distance of the}$$

C. G. of the shell from the base.

(6) Let length of beam be $6a$; then height will be $3a$, and length of string will be $4a$; evidently 1 ft. of beam is resting on wall. Take moments about intersection of reactions and tension of string is found to be 14.4 lbs.

(7). Since AB = 13 and AD = 5, BD = 12; also BC = $3\sqrt{41}$. Through A draw AE parallel to BD; then AE : BD :: CA : CD; therefore AE = 8. Also, EB : EC :: 5 : 10; therefore, EB = $\sqrt{41}$. The sides of the triangle ABE are parallel to the three forces which keep the point C at rest, and are therefore proportional to them. Hence,

tension of BC : $13\frac{1}{2}$ :: $\sqrt{41}$: 8

$$\therefore \text{ tension of BC} = \tfrac{27}{16}\sqrt{41} = 10.805 \text{ cwt.}$$

Again, Pressure on spar : $13\frac{1}{2}$ :: 13 : 8

$$\therefore \text{ Pressure on spar} = \frac{13\frac{1}{2} \times 13}{8} = 21.938 \text{ cwt.}$$

Or thus: From A draw AF perpendicular to AB. Then AF $\times 3\sqrt{41} = 10 \times 12$, each being double triangle ABC.

$\therefore$ AF $= \dfrac{40}{\sqrt{41}}$. Let $t =$ tension of stay. Take moments about A, and $t \times \dfrac{40}{\sqrt{41}} = 13\tfrac{1}{2} \times 5$; $\therefore t = \dfrac{27\sqrt{41}}{16}$, as before.

From C draw CG perpendicular to BA produced. Then $13 \times CG = 10 \times 12$, or $CG = \tfrac{120}{13}$. Let p be pressure on spar. Take moment about C, and $p \times \tfrac{120}{13} = 13\tfrac{1}{2} \times 15$, or $p = \dfrac{13\tfrac{1}{2} \times 13}{8}$, as before.

It will also be a profitable exercise to solve by resolving vertically and horizontally.

(8) 6.276 cwt.; 37.4192.

(9) Let l, l_1 be the lengths of the two planes. Let P, Q, be the two weights at a distance from the common vertex of the planes $= x$ and y respectively, where $x + y = c$, say. Through vertex of planes draw line parallel to base; take moments about this line, and $(P + Q)x = Pp + Qq$; where p and q are perpendiculars from weights on parallel line through vertex.

If h be height of planes, then $\dfrac{p}{x} = \dfrac{h}{l}$ $\therefore p = \dfrac{hx}{l}$, and $q = \dfrac{hy}{l_1}$. Resolve both weights along the plane, and $P\dfrac{h}{l} = Q\dfrac{h}{l_1}$. Hence we have

$$(P + Q)x = P\dfrac{hx}{l} + Q\dfrac{hy}{l_1} = P\dfrac{hx}{l} + P\dfrac{hy}{l}$$
$$= P\dfrac{h}{l}(x + y) = P\dfrac{hc}{l} = \text{constant}.$$

$\therefore x =$ constant; that is, the centre of gravity is independent of the position of the weights, and is always at a constant distance from the parallel line through the vertex, and must therefore be in a straight line.

(10) $7\tfrac{2}{9}$ feet.

Chapter XII

EXAMINATION PAPERS.

Special Examination for Inspectors. Page 146.

(1) 0. (2) Art. 105; P : W :: 1 : 8. (3) Arts. 121, 123, 132. (4) In the first case we find $2P = W$; in second case we find the Power equal to the Weight. ∴ $2P$ would be required to sustain W.

(5) The following clause of this question has been omitted:—"And the other end is sustained by a force of 40lbs. acting at right angles to the length of the beam, and in a vertical plane passing through the beam."

Let a = length of beam, and x = distance of lower end of beam from direction of wt. Take moments about foot of ladder, and $40a = 100x$, or $x = \frac{2}{5}a$. ∴ base of right-angled triangle formed by the ladder is $\frac{2}{5}a$, and height will be $\frac{3}{5}a$. Hence sides of right-angled triangle are in the ratio of 3, 4, 5.

Produce reaction of ground till it meets direction of force 40lbs. If x be distance from pt. of intersection to horizontal plane, by similar triangles $\dfrac{x}{5} = \frac{5}{3}$, ∴ $x = \frac{25}{3}$.

Take moments about intersection of reaction and direction of 40, and $F \times \frac{25}{3} = 100 \times 2$ ∴ $F = 24$. Take moments about top of ladder,
and $R \times 4 = 24 \times 3 + 100 \times 2$.
∴ $R = 68$.

Second Class Certificates. Page 147.

(1) 4lbs. and 6lbs. (2) Art. 124. (3) The Centre of Gravity is at the point B. (4) a 26, b 27. (5) 60ft. of height for 100ft. of length.

(6) (a) Weight of beam, acting vertically downwards at its middle point; the pressure exerted by the man acting as described in the question; the reaction of the ground against the end of the beam which rests on the ground, acting vertically upwards; and friction, acting horizontally. (b) 50lbs.

First Class, July 1871. Page 148.

(1) (*a*) Art. 123. (*b*) 131. (2) 5 at A, 4 at B, 6 at D, 5 at E. (3) $4\frac{4}{75}$lbs. (4) 47.

Second Class, December 1871. Page 148.

(1) Arts. 97 and 99. (2) Arts. 118 and 119. (3) Art. 132. (4) The forces are equivalent to 2AC. The required line will be equal to 2AC and a direction CA. (5) $2\frac{1}{2}$lbs.

First Class, December 1871. Page 149.

(1) Art. 128. (2) $11\frac{1}{2}$lbs.

Normal School, 1871.

(1) 40lbs. (2) 100lbs. (3) $145\frac{5}{7}$lbs. (4) Arts. 66 and 69. (5) On CA, 525lbs. and on DB, 475lbs.

Second Class, July 1872. Page 150.

(1) Art. 88. (2) Take moments about fulcrum and force $= \dfrac{1000\sqrt{21}}{21}$. (3) Art. 118. (4) 80lbs.

(5) Length of line representing resolved parts is 12ft., and magnitude of resultant is 15ft.

(6) Art. 136. The friction is independent of the velocity when there is sliding motion.

First Class, July 1872. Page 151.

(1) From C draw CE, CF, perpendiculars on AD, BD respectively. Since the lever is at rest, the moments about C must be equal in magnitude and opposite in direction; that is
$$AD.CE = BD.CF;$$
$$\therefore \text{triangle } ACD = \text{triangle } BCD;$$
$$\therefore AC = CB.$$

(2) Resolve each force vertically and horizontally. The horizontal components are 1, 1, acting in opposite directions and, therefore, neutralizing each other. The sum of the vertical components is the resultant and is at right angles to AB.

Second Class, July 1873. Page 151.

(1) Art. 88. (2) (*a*) Friction, art. 135. (*b*) Art. 136.
(3) Art. 107. (4) Arts. 121, 123, and 129.
(5) Arts. 134, 135, and 137.

First Class, July 1873. Page 152.

(1) Resolve the forces vertically and horizontally.

(2) See note to question 11, page 40, and question 12, page 115; $P = 62\frac{1}{2}$ ℔s.

(3) Let $62\frac{1}{2}$ be at A, and suppose it drawn up to B; then 100 will descend to a point D, 25ft. from B. Through D draw DE parallel to AC. Then

$$BE : 25 :: h : 40$$
$$\text{or } BE = \tfrac{5}{8}h.$$

The Principle of Virtual Velocities asserts that $62\frac{1}{2} \times h = 100 \times \tfrac{5}{8}h$, which is true, and therefore the Principle holds in this case.

(4) Ex. 3., page 142. AD should be BD.

(5) Resolve vertically and horizontally, and we have

$$1\tfrac{2}{13}t + 1\tfrac{2}{13}t' = 112.$$
and $$\tfrac{9}{13}t = \tfrac{5}{13}t'. \quad \text{find } t'.$$

Take moments about A, and

$$56 \times 7 + 56 \times x = 1\tfrac{2}{13}t' \times 14.$$

$\therefore x = 11 = AE$, whence $BE = 3$.

Or, we may obtain the same result by the Triangle of Forces. Resolve the weight of the rod into 28 at A and 28 at B. Also resolve the 56 acting at a distance x from A into $56 - 4x$ at A and $4x$ at B. Then proceeding as in Ex. 4, page 23, we have

$$\frac{84 - 4x}{T} = \frac{12}{9},$$
and $$\frac{T}{28 + 4x} = \frac{5}{12};$$

$\therefore x = 11$, as before.

Normal School, June 1873. Page 153.

(1) $100\sqrt{3}$. (2) 25 ℔s. (3) Arts. 66 and 77. Take an obtuse-angled triangle, vertex B, obtuse angle C, and acute angle A. Place so that AC may be vertical, A being above C. Let ACD be another obtuse-angled triangle similarly situated their sides AC coinciding. Then ABCD will be a quadrilateral. Join BD and produce AC to meet BD in E. Let $AE = x$; $CE = y$. The areas ABD, BCD, ABCD are as x, y, and $x-y$ and the distances of their centre of gravity from BD, are as

$$\frac{x}{3}, \frac{y}{3} \text{ and } y:$$

$$\therefore x \cdot \frac{x}{3} = y \cdot \frac{y}{3} + (x-y)y; \therefore \frac{x^2 - y^2}{3} = (x-y)y:$$

and since x and y cannot be equal

$$\frac{x+y}{3} = y, \text{ and } \frac{x}{y} = 2.$$

(4) 120 ℔s. (5) 60°. (6) 8 ℔s.

Second Class, December 1873.

(1) Arts. 97, 98, 99. (2) Art. 94; 16 oz. (3) The tension is equal to the weight. (5) (*a*) Art. 105. (*b*) 4. (6) The sum of the resolved parts in the direction AB is $\tfrac{3}{4}$; sum of the resolved parts in direction AD is also $\tfrac{3}{4}$. Therefore resultant $= \tfrac{3}{4}\sqrt{2}$.

First Class, December 1873. Page 155.

(1) This question has been misprinted. After respectively, read—" and in direction by AD, EA (not AE), and AB, respectively, prove that their resultant is in the direction of the diagonal of a square described on AC."

The forces resolved along $AC = \tfrac{AB}{4}(1 - \sqrt{3})$.
The forces resolved at right angles to $AC = \tfrac{AB}{4}(1 - \sqrt{3})$

The resultant is, therefore, in direction of diagonal of square described on AC.

(2) Let reaction of plane meet AB in F; then direction of wt. of rod will pass through point of intersection. Through C draw CG, parallel to AB, meeting direction of weight in G. Triangle CFG has its sides parallel to the three forces which keep rod at rest, and are, therefore, proportional to them. Hence $t : R :: GC : CF$, but $GC = CF \therefore t = R$.

(3) 1lb. at A.

Second Class, July 1874. Page 156.

(1) 40lbs. (2) (*a*) Art. 123. (*b*) 125. (3) (*a*) Art. 105. (*b*) 128. (4) (*a*) Art. 23. (*b*) 12 feet.

First Class, July 1874. Page 156.

(1) There should be 20lbs. acting at B. $m = 45$lbs. and $n = 5\sqrt{3}$. (2) Resolve along the plane and at right angles to it.

Second Class, July 1875. Page 157.

(1) $P = 14$lbs. and $W = 10$lbs. (2) Art. 132. (3) It acts at a mechanical disadvantage when the height is greater than the base.

(4) The reaction of the plane on the body, exerted in a direction at right angles to the plane. (5) Arts. 19 and 36. (6) $\frac{1}{2}$ft. and $\frac{1}{2}\sqrt{3}$ ft.

First Class, July 1875. Page 158.

Resolve vertically and horizontally, and we have $m : n :: \sqrt{3} - 1 : 2\sqrt{2}$.

(2) $h : b :: 1 : 2$ (3) In the parenthesis at the end of this question, read ADC instead of ACD. A vertical force of $1\frac{2}{3}$lbs. acting upwards at A and an equal force acting downwards at C, will produce equilibrium

3. In order that the rod may be in equilibrium, its weight, supposed to be collected at its middle point O, must lie vertically below D, the intersection of the reactions of the planes at F and G: thus the semi-diagonal

OD of the rectangle BFDG and therefore the whole diagonal BD must be vertical. The triangle BOF is, therefore, equilateral, and BF = OF = half length of rod.

First Class, 1875. Page 159.

(1) The third sentence should read: "Find the magnitude of the resultant of four forces, &c." The resultant of the forces acting along the sides of the square $= 3\sqrt{2}$. The resultant of forces, each $= \sqrt{6}$, acting at an angle of $60°= 3\sqrt{2}$; and these resultants act in opposite directions. Therefore $R = 0$.

(2) Resolve vertically and horizontally and take moments about A.

(3) Resolve vertically and horizontally and take moments about A.

Second Class, 1875. Page 160.

(2) Show that moments about fulcrum are equal for every position.

(3) Take moments about E. 1 inch from D towards B. (4) Art. 29. (5) $\frac{5}{13}$ of the whole force. (6) (*b*) When there is no mechanical advantage $W = P$. In this case $\dfrac{P}{4} + \dfrac{Q}{4} + \dfrac{Q}{2} = P \therefore \dfrac{P}{Q} = 1$.

Second Class, July 1876. Page 161.

(1) Art. 10. If the three forces pass through a point, they are capable of being represented by the sides of a triangle. If they do not pass through a point they are parallel, then the algebraic sum of their moments about any point must vanish. (2) $5\sqrt{3}$ lbs. The line of action of the resultant will be perpendicular to that of the 1 lb. force, and will, therefore, be equally inclined to the lines of action of the 6 lbs. and 4 lbs. forces.

(3) Arts. 52 and 58. (*b*) 2.40 lbs. (4) See Ex. 4, page 84. $1\frac{1}{2}$ inches from centre.

(5) The volume is proportional to the weight and may, therefore, be used for it. Take moments about base using volume for weight, and we have

$$\frac{h}{2} \cdot \frac{r_1^2 + 3r_2^2 + 5r_3^2 + \ldots (2n-1)r_n^2}{r_1^2 + r_2^2 + r_3^2 + \ldots \ldots + r_n^2}.$$

First Class, July 1876. Page 161.

(1) The enunciations of the Triangle of Forces given in articles 29 and 32 may be still further extended as follows :—If three forces, the magnitudes of which are proportional to the sides BC, CA, AB, of a triangle ABC, and of which the directions are either parallel to these sides or respectively inclined to them at equal angles, act upon a point they will produce equilibrium, and conversely.

Three forces acting in consecutive directions round a triangle ABC (AC horizontal) are equivalent to a couple (Art. 49). To show this, through A draw AD equal and parallel to CB, and in it introduce a pair of balancing forces, each equal to CB. Of the five forces, three AC, AD and BA, are in equilibrium, and may be removed; there are then left two forces CB and DA, equal, parallel, and in contrary directions which constitute a couple.

(2) The direction of the weight produced must pass through the fulcrum (art. 59). The question then becomes a simple geometrical one, viz., having given the three sides of a triangle to find the segments made by a perpendicular from the vertex on the base. If x and y be the segments, then

$$a^2 - b^2 = x^2 - y^2 = (x+y)(x-y) = c(x-y)$$

$$\therefore x - y = \frac{a^2 - b^2}{c}$$

and $x + y = c$

$$\therefore x = \frac{a^2 + c^2 - b^2}{2c},$$

and $\qquad y = \dfrac{b^2 + c^2 - a^2}{2c}$;

$$\therefore x : y :: a^2 + c^2 - b^2 : b^2 + c^2 - a^2.$$

(3) Art. 81. (4) Applying the Triangle of Forces, we find strain on hinge 146lbs. and pressure on point 110lbs. In the second case the strain and pressure will be 349lbs. and 272lbs. respectively.

Intermediate Examination, June 1876.
Page 162.

(1) Art. 23. (2) Apply the Triangle of Forces. See note on question 2, Exercise III, page 27. Force required $= \dfrac{10\sqrt{7}}{3}$ cwt. (3) Art. 52. 50lbs. (4) 20lbs. (5) Arts. 121, 123 and 131.

Intermediate Examination, December 1876.
Page 163.

(1) 40lbs. (2) See Ex. 2, page 59. As 1 : 2.
(3) (*a*) Art. 23. (*b*) G $= \tfrac{1}{2}\sqrt{3}$. (4) Arts. (*a*) 121 and 123; (*b*) 107; (*c*) 129.

www.ingramcontent.com/pod-product-compliance
Lightning Source LLC
Chambersburg PA
CBHW021728220426
43662CB00008B/759